# Aquatic Macrophyte Risk Assessment for Pesticides

T0179050

# Other Titles from the Society of Environmental Toxicology and Chemistry (SETAC)

*Veterinary Medicines in the Environment*
Crane, Boxall, Barrett
2008

*Relevance of Ambient Water Quality Criteria for Ephemeral and Effluent-
dependent Watercourses of the Arid Western United States*
Gensemer, Meyerhof, Ramage, Curley
2008

*Extrapolation Practice for Ecotoxicological Effect Characterization of Chemicals*
Solomon, Brock, de Zwart, Dyev, Posthumm, Richards, editors
2008

*Environmental Life Cycle Costing*
Hunkeler, Lichtenvort, Rebitzer, editors
2008

*Valuation of Ecological Resources: Integration of Ecology and Socioeconomics
in Environmental Decision Making*
Stahl, Kapustka, Munns, Bruins, editors
2007

*Genomics in Regulatory Ecotoxicology: Applications and Challenges*
Ankley, Miracle, Perkins, Daston, editors
2007

*Population-Level Ecological Risk Assessment*
Barnthouse, Munns, Sorensen, editors
2007

*Effects of Water Chemistry on Bioavailability and Toxicity of Waterborne Cadmium,
Copper, Nickel, Lead, and Zinc on Freshwater Organisms*
Meyer, Clearwater, Doser, Rogaczewski, Hansen
2007

*Ecosystem Responses to Mercury Contamination: Indicators of Change*
Harris, Krabbenhoft, Mason, Murray, Reash, Saltman, editors
2007

For information about SETAC publications, including SETAC's international journals, Environmental Toxicology and Chemistry and Integrated Environmental Assessment and Management, contact the SETAC Administratice Office nearest you:

SETAC Office
1010 North 12th Avenue
Pensacola, FL 32501-3367 USA
T 850 469 1500   F 850 469 9778
E setac@setac.org

SETAC Office
Avenue de la Toison d'Or 67
B-1060 Brussells, Belguim
T 32 2 772 72 81   F 32 2 770 53 86
E setac@setaceu.org

***www.setac.org***
**Environmental Quality Through Science®**

# Aquatic Macrophyte Risk Assessment for Pesticides

Lorraine Maltby   Dave Arnold   Gertie Arts

Jo Davies   Fred Heimbach   Christina Pickl

Véronique Poulsen

SETAC Europe Workshop AMRAP
Wageningen, Netherlands

Coordinating Editor of SETAC Books
Joseph W. Gorsuch
Gorsuch Environmental Management Services, Inc.
Webster, New York, USA

SETAC

CRC Press
Taylor & Francis Group
Boca Raton London New York

CRC Press is an imprint of the
Taylor & Francis Group, an **informa** business

Information contained herein does not necessarily reflect the policy or views of the Society of Environmental Toxicology and Chemistry (SETAC). Mention of commercial or noncommercial products and services does not imply endorsement or affiliation by the author or SETAC.

CRC Press
Taylor & Francis Group
6000 Broken Sound Parkway NW, Suite 300
Boca Raton, FL 33487-2742

First issued in paperback 2019

© 2010 by Taylor & Francis Group, LLC
CRC Press is an imprint of Taylor & Francis Group, an Informa business

No claim to original U.S. Government works

ISBN-13: 978-1-4398-2211-1 (hbk)
ISBN-13: 978-0-367-38492-0 (pbk)

**Library of Congress Cataloging-in-Publication Data**

Aquatic macrophyte risk assessment for pesticides / Lorraine Maltby ... [et al.].
    p. cm.
  Includes bibliographical references and index.
  ISBN 978-1-4398-2211-1 (hardcover : alk. paper)
    1. Freshwater plants--Effect of pesticides on. 2. Pesticides--Toxicity testing. I. Maltby, Lorraine.

QK105.A646 2010
581.7'6--dc22                                                                  2009031797

Visit the Taylor & Francis Web site at
http://www.taylorandfrancis.com

and the CRC Press Web site at
http://www.crcpress.com

# SETAC Publications

Books published by the Society of Environmental Toxicology and Chemistry (SETAC) provide in-depth reviews and critical appraisals on scientific subjects relevant to understanding the impacts of chemicals and technology on the environment. The books explore topics reviewed and recommended by the Publications Advisory Council and approved by the SETAC North America, Latin America, or Asia/Pacific Board of Directors; the SETAC Europe Council; or the SETAC World Council for their importance, timeliness, and contribution to multidisciplinary approaches to solving environmental problems. The diversity and breadth of subjects covered in the series reflect the wide range of disciplines encompassed by environmental toxicology, environmental chemistry, hazard and risk assessment, and life-cycle assessment. SETAC books attempt to present the reader with authoritative coverage of the literature, as well as paradigms, methodologies, and controversies; research needs; and new developments specific to the featured topics. The books are generally peer reviewed for SETAC by acknowledged experts.

SETAC publications, which include Technical Issue Papers (TIPs), workshop summaries, newsletter (*SETAC Globe*), and journals (*Environmental Toxicology and Chemistry* and *Integrated Environmental Assessment and Management*), are useful to environmental scientists in research, research management, chemical manufacturing and regulation, risk assessment, and education, as well as to students considering or preparing for careers in these areas. The publications provide information for keeping abreast of recent developments in familiar subject areas and for rapid introduction to principles and approaches in new subject areas.

SETAC recognizes and thanks the past coordinating editors of SETAC books:

> A.S. Green, International Zinc Association
>    Durham, North Carolina, USA
> C.G. Ingersoll, Columbia Environmental Research Center
>    US Geological Survey, Columbia, Missouri, USA
> T.W. La Point, Institute of Applied Sciences
>    University of North Texas, Denton, Texas, USA
> B.T. Walton, US Environmental Protection Agency
>    Research Triangle Park, North Carolina, USA
> C.H. Ward, Department of Environmental Sciences and Engineering
>    Rice University, Houston, Texas, USA

# Contents

# List of Figures

# List of Tables

# About the Authors

**Lorraine Maltby** is Professor of Environmental Biology and Head of the Department of Animal and Plant Sciences at The University of Sheffield, UK. She has 25 years' experience in the fields of freshwater ecology and ecotoxicology and has published more than 100 scientific articles. Her research aims to understand how ecosystems respond and adapt to environmental stressors, including pollutants. This understanding is fundamental to the successful protection and management of ecosystems in order to ensure the sustainable provision of the ecosystem goods and services that underpin human well-being. Maltby has served on UK and international scientific advisory bodies on pesticides and other chemical substances and has been an active member of the Society of Environmental Toxicology and Chemistry (SETAC) since 1989, having served as the President of SETAC Europe in 2001 and the founding President of the global SETAC organization in 2002. She has organized and participated in a number of SETAC workshops, annual meetings, and World Congresses over the last 20 years.

**Dave Arnold** is an environmental consultant with more than 35 years' experience in industry and consultancy focusing on pesticide fate and behavior, ecotoxicology, and risk assessment. He was chair of the European Crop Protection Association (ECPA) Environmental Exposure Assessment Group and, as a member of the EU FOCUS Steering Committee in DG SANCO (Consumer Health and Protection), was involved in the management of the development of EU FOCUS soil, climate, cropping scenarios for modeling pesticide risks to ground and surface water under Directive 91/414. He was involved in the development of OECD guidelines for the testing of chemicals including soil microbiology, earthworm toxicity, and fate in sediment and water. He is an active member of SETAC since the formation of SETAC Europe in 1989. He was Treasurer and President of both the UK Branch of SETAC Europe and SETAC Europe.

He has been a member of the organizing committee of a number of SETAC workshops contributing to their ensuing publications, including the SETAC document on testing procedures in freshwater mesocosms (1991) and Higher-tier Aquatic Risk Assessment for Pesticides (HARAP) in 1998. More recently he was a member of the SETAC Europe ELINK workshop (linking exposure and effects of pesticides in the aquatic environment).

**Gertie Arts** is Doctor in Biology and works at Alterra as a senior scientist in Environmental Risk Assessment. She is involved in higher-tier experiments for higher-tier aquatic risk assessment procedures for contaminants (e.g., for the registration of pesticides). She works on the ecological evaluation of pesticide risks and on ecological and other aspects of freshwaters in the Dutch agricultural landscape, for example, multistress and eutrophication. Recently, she is more involved in aquatic macrophyte risk assessment, performing and directing research in the laboratory and in mesocosms. Her research focuses on the effects of contaminants on aquatic macrophytes in different experimental settings and at different levels of biological organization. Within the Dutch Pesticides Research Program, she participates as a project representative of the theme "Ecological Risk Assessment of Pesticides," dealing with the effects and ecological risks in surface water. She has been a member of SETAC 2006 Scientific Committee. She is co-chair of the Organizing Committee of the AMRAP workshop focusing on aquatic macrophyte risk assessment for plant protection products, and she is chair of the Steering Committee of the SETAC Aquatic Macrophytes Ecotoxicology Group (AMEG).

**Jo Davies** is an environmental safety specialist with Syngenta at Jealott's Hill International Research Centre in the UK. She obtained a degree in agricultural botany from the University of Reading and a PhD in biochemistry from the University of Bristol. During this time, she specialized in weed science, particularly herbicide and safener mode of action and metabolism. Davies has since worked on a diverse range of pesticide-related topics including the effects of environmental factors on herbicide activity and the development of aquatic and terrestrial bioassays to detect herbicide activity as well as working as an analytical chemist on regulatory studies. In 1997, she began work on a DEFRA-sponsored project to develop ecotoxicity tests for assessing herbicide effects on non-target aquatic plants. She has subsequently conducted many laboratory and field-based herbicide efficacy and ecotoxicity tests with aquatic plants. Davies joined Syngenta in 2001, where she is responsible for the environmental safety assessments for several products and provides expertise in risk assessment for aquatic plants. She is also a member of the steering committee of the SETAC Aquatic Macrophyte Ecotoxicology Group (AMEG).

**Fred Heimbach** works as a consultant scientist at RIFCon GmbH in Leichlingen, Germany. He obtained his MSc degree and PhD in conducted research on marine insects at the Institute of Zoology, Physiological Ecology at the University of Cologne.

From 1979 until 2007 he worked at Bayer CropScience in Monheim, Germany on the side effects of pesticides on non-target organisms. In addition to his work, he gave lectures on ecotoxicology at the University of Cologne. Dr. Heimbach has researched the development of single-species toxicity tests for both terrestrial and aquatic organisms and has worked with microcosms and mesocosms in the development of multispecies tests for these organisms. As an active member of European and international working groups, he participated in the development of suitable test methods and risk assessment of pesticides and other chemicals for their potential side-effects on non-target organisms.

For several years he has served on the SETAC Europe Council and the SETAC World Council, and he has been an active member of the organizing committees of several European workshops on specific aspects of the ecotoxicology of pesticides.

**Christina Pickl** holds a diploma in Agricultural Biology from the University of Stuttgart-Hohenheim, Germany, and a PhD in Ecotoxicology and Environmental Chemistry. She was mainly working on the development of biological test systems for environmental samples with specific methodological requirements, as for example, the assessment of acidic mining lakes. After graduation she was Managing Director of ÖkoTox GmbH, a start-up company for biological test systems. During that time she was involved in the method development of different macrophyte tests and other plant tests, including the establishment of stock cultures and computerized image analysis systems. In 2004 she joined the German Federal Environment Agency (UBA) and worked in the section "Plant Protection Products" responsible for the environmental risk assessment and management. Currently she is Deputy Head of the section "Plant Protection Products" and within the section leader of the subgroup "Exposure, Degradation and Groundwater Risk Assessment and Management."

**Véronique Poulsen** is a senior ecotoxicological and environmental risk assessor. She is Head Deputy of the Ecotoxicological and Environmental Fate Unit in the Plant and Environment Department at the French Food Safety Agency (AFSSA). She obtained her PhD in conducted research on aquatic microcosms at the National Institute of Industrial Environment and Risks (INERIS), France.

She was in charge of applied research programs on aquatic mesocosms and of methodological development of ecotoxicological laboratory single-species tests at INERIS from 1996 to 2005. She has been involved in risk assessment for several years, starting with chemicals and biocides from 2002 to 2006 at INERIS/BERPC, before joining AFSSA in November 2006 to examine environmental risk assessment for pesticide dossiers.

Poulsen was involved in several European projects during her career: SedNet (Sediment Network) where ecotoxicological methods to be used to evaluate the toxicity of sediments in support of the European Water Framework Directive were developed; NORMAN co-ordination action (network of reference laboratories and related organizations for monitoring and biomonitoring of emerging environmental pollutants) under the 6th European Framework Program.

# Executive Summary

This publication is the output from the Society of Environmental Toxicology and Chemistry (SETAC) Europe workshop on Aquatic Macrophyte Risk Assessment for Pesticides (AMRAP) held in the Netherlands in January 2008, which was attended by scientists from regulatory authorities, business, and academia. The workshop was initiated in response to concerns by the scientific and regulatory community that the current risk assessment scheme for plant protection products in the European Union (EU) may not provide adequate protection for aquatic macrophytes. There is clear scientific evidence to support the contention that aquatic macrophytes play a key role in the structure and functioning of aquatic ecosystems and, hence, must be considered within the risk assessment process. Current data requirements under Council Directive 91/414/EEC provide information on the toxicity of herbicides and plant growth regulators (PGRs) to algae and *Lemna* (EU 1997). Council Directive 91/414/EEC is currently being revised and the EU Guidance Document on Aquatic Ecotoxicology (EC 2002) will be revised over the next few years. The issues discussed in this workshop are pertinent to these revisions and should be considered accordingly. However, there is a clear requirement to build regulatory confidence in the application of new methods for aquatic macrophyte risk assessment and it is partly for that reason that this AMRAP document has been prepared.

Key outputs from the AMRAP workshop are consolidated in Chapter 2 in the form of guidance for an improved approach to aquatic macrophyte risk assessment. Outputs include a proposed decision scheme for assessing the risk of herbicides and PGRs to aquatic macrophytes and a series of 12 recommendations that were formulated from the workshop discussions. The background and rationale behind each recommendation and point in the proposed decision scheme are also documented. The guidance and recommendations are distilled from existing regulatory experiences of aquatic macrophyte risk assessment; the interrogation of case studies to identify issues, data gaps, and inadequacies in study design; and the outputs from plenary discussions that identified improvements to risk assessment that could be implemented immediately and those for which further research is needed.

The key regulatory concern was that risk assessments based on *Lemna* and algal data only at Tier 1 may underestimate the risk of plant protection products to aquatic macrophytes. In particular, concern was raised that *Lemna* and algae may not be sensitive to some herbicides that form residues in sediment or have modes of action that are not expressed in *Lemna*. The risk assessment scheme for aquatic macrophytes proposes that where assessment criteria indicate concern, then a rooted macrophyte species should be tested. Because of considerable knowledge and experience with *Myriophyllum*, this species is recommended as the additional Tier 1 test species.

If the Tier 1 level of concern is exceeded, then higher-tier risk assessments are recommended. Higher-tier assessments may take the form of either mitigation of exposure to refine the predicted environmental concentration or the refinement of effects, either through additional tests with modified exposure regimes, the

1

generation of further single-species data for use in species sensitivity distributions (SSDs), or by conducting multispecies, microcosm or mesocosm tests. Guidance and recommendations are provided on each of these approaches.

## RECOMMENDATIONS FOR ASSESSING THE RISKS OF HERBICIDES AND PLANT GROWTH REGULATORS TO AQUATIC MACROPHYTES

1) Conduct an additional test with a rooted macrophyte species when either *Lemna* is known not to be sensitive to the test compound's mode of action (MoA), or there is a lack of expected herbicidal activity in Tier 1 *Lemna* and algal tests, or where exposure via sediment may be a critical factor in the risk assessment.

2) Assess the effectiveness and reproducibility of an agreed test protocol for a rooted macrophyte (*Myriophyllum* sp.) via a ring test.

3) Assess growth in additional aquatic macrophyte tests using biomass and shoot length measurements.

4) Consider the exposure profile in relation to the species and effect under investigation, consider the length of the study required in relation to the expected exposure profile, and take into account the ecological context of the scenarios under scrutiny when higher-tier studies are designed using modified exposure regimes.

5) Collate a list of aquatic macrophyte species to assist the selection of appropriate species for evaluation in higher-tier single-species, multispecies, microcosm, and mesocosm tests.

6) Collate and analyze data on single-species macrophyte toxicity to enable an assessment of the relative sensitivity of *Lemna* and other macrophyte species.

7) Include a range of morphologically and taxonomically different macrophytes in SSDs. Where feasible, endpoints should be based on a common measurement for all species.

8) Disseminate AMRAP guidance concerning the construction and use of SSDs for aquatic macrophytes with the aim of reaching agreement on SSD criteria and outputs for use within the regulatory framework.

9) Ensure that mesocosm studies are appropriately designed to answer questions concerning effects on sensitive species, for example, using a potted plant approach; effects (direct and indirect) on natural (established) communities; or a combination of both approaches. Mesocosm studies addressing the risks of herbicides or PGRs should contain a sufficient variety of morphological forms and taxonomic groups to enable an adequate assessment of risk.

10) Include *Lemna* in mesocosm studies where feasible and appropriate. Where conditions for the growth of submerged macrophytes are not optimal for *Lemna*, separate bioassays or other higher-tier experiments using *Lemna* may be used.

11) Develop tools for the temporal–spatial extrapolation of mesocosm data in order to gain a better understanding of the ability of mesocosms to reflect macrophyte responses in natural systems.
12) Establish an aquatic macrophyte advisory group under the auspices of SETAC to continue the development of risk assessment tools and to steer education and training in aquatic macrophyte ecotoxicology.

Four workgroups were established to generate information to support Tier 1 and higher-tier risk assessment:

1) the development and validation of decision-making criteria to underpin the proposed decision scheme for Tier 1;
2) the development and ring-testing of a study design using either *Myriophyllum spicatum* or *Myriophyllum aquaticum*, intended as the additional species test at Tier 1;
3) collation of a database of macrophyte species and test methods based on current experience; and
4) development of criteria and guidance on species and endpoints and their use in SSDs.

It is hoped that the information provided through this publication will assist in the development of an improved assessment scheme for evaluating the risk of plant protection products to aquatic macrophytes and that the AMRAP workgroups will enable ongoing scientific debate among all stakeholders. The ongoing activities of the workgroups will be facilitated by a SETAC Advisory Group, Aquatic Macrophyte Ecotoxicology Group (AMEG). This group will act as a focal point for ongoing discussion and development of the science of risk assessment for aquatic macrophytes. The inaugural meeting of AMEG was held in June 2009 in Göteborg, Sweden.

# 1 Introduction and Background

All life on earth depends directly or indirectly on primary production, and primary producers play a key role in the structure and functioning of aquatic ecosystems. Primary producers influence the chemical status of waters, produce the oxygen required by aquatic biota, provide food for herbivores and detritivores, enhance habitat complexity, and provide substrate and shelter for a diversity of other plants and animals.

Freshwater primary producers can be divided into microscopic algae, including photosynthetic bacteria, and aquatic macrophytes. Aquatic macrophytes are a diverse assemblage of plants that have become adapted for life wholly or partially in water. They are photosynthetic organisms that are large enough to be seen with the naked eye and include bryophytes (mosses), pterophytes (ferns), equisetophytes (horsetails), and magnoliophytes (flowering plants), as well as macroalgae such as Charophyceae (e.g., *Chara* and *Nitella*) and Ulvophyceae (e.g., *Enteromorpha*). Flowering plants are the most obvious group of aquatic macrophytes, and both monocotyledons (Liliopsida) and dicotyledons (Magnoliopsida) occur in freshwaters.

Aquatic macrophytes are often classified by their growth habit, the four categories being emergent, rooted and floating-leaved, free-floating, and submerged (see Figure 1.1).

Emergent macrophytes are rooted in the substratum, with most of their leaves and flowers above the water surface (e.g., *Glyceria maxima*, *Typha latifolia*, *Phragmites australis*). Floating-leaved macrophytes are also rooted in the substratum, but most of their leaf tissue is at the water surface (e.g., *Nymphaea alba*, *Potamogeton natans*). Free-floating macrophytes are not rooted and float unattached either on the water surface (e.g., *Lemna minor*, *Hydrocharis morsus-ranae*) or in the water column (e.g., *Ceratophyllum demersum*, *Utricularia vulgaris*). Submerged macrophytes are rooted in the substratum with most of their vegetative tissue below the water surface (e.g., *Myriophyllum spictatum*, *Elodea canadensis*).

Given the essential role that primary producers play in aquatic ecosystems, it is imperative that the potential risk of pesticides to the structure and functioning of aquatic plants is adequately assessed. Council Directive 91/414/EEC (EU 1997) sets out the risk assessment framework for pesticides used to protect plants or plant products against harmful organisms. These pesticides are classified as plant protection products, and the specific data requirements, including information on toxicity to aquatic plants, are given in Annexes II and III of the Directive. Current risk assessment procedures require that all plant protection products are tested against a green algae and that herbicides and plant growth regulators (PGRs) are also tested against a second algal species (from a different taxonomic group) and an aquatic macrophyte species. The Guidance Document SANCO 3268/2001 on Aquatic Ecotoxicology

## Habitat

submerged            floating-leaved            free-floating            emergent

**FIGURE 1.1**   Growth habit of aquatic macrophytes.

states that macrophyte tests should be conducted with the duckweed *Lemna* and that additional data with other plant species may be required on a case-by-case basis (EC 2002). *Lemna* is a non-sediment-rooted monocotyledon with a short generation time. The scientific and regulatory communities have raised concerns that risk assessments based on *Lemna* toxicity may not be protective of other macrophyte species due to potential differences in exposure route, recovery rate, or sensitivity to specific toxic modes of action (Brock et al. 2000; Vervliet-Scheebaum et al. 2006).

To address these issues, the workshop Aquatic Macrophyte Risk Assessment for Pesticides (AMRAP) was held under the auspices of SETAC Europe in Wageningen, Netherlands, 14–16 January 2008. The workshop was attended by 41 participants from 10 countries (Appendix II). In keeping with SETAC philosophy, representation was tripartite with 29% of participants from academia, 34% from government, and 37% from business.

The aim of the workshop was to synthesize current knowledge in order to provide guidance for the use and interpretation of non-target aquatic macrophyte data in the risk assessment of plant protection products in Europe. This aim was addressed by

- presenting an overview of the current European regulatory framework for the risk assessment of aquatic macrophytes,
- identifying uncertainties and areas for reducing uncertainties within the regulatory framework,
- presenting and discussing the current state of the science of aquatic macrophyte testing in single-species laboratory studies and mesocosm studies,
- evaluating the extent to which currently available methods and understanding can address the uncertainties in the risk assessment of aquatic macrophytes, and
- making recommendations for improving aquatic macrophyte testing methodologies and risk assessment.

At the workshop, keynote presentations considered the diversity and importance of macrophytes in agricultural landscapes, laboratory and field methods for macrophyte studies, and the current European regulatory framework for risk assessment. During several plenary sessions and case study discussions, participants were asked to identify areas of uncertainty within the regulatory framework and to discuss the strengths and limitations of existing test methods for aquatic macrophytes. A number of areas of uncertainty were identified, and workgroups were established to develop recommendations for each of these areas, summarized in brief below.

Workgroup 1: Develop decision-making criteria to determine when *Lemna* may not be an appropriate test species. Chair: Eric Bruns.

Workgroup 2: Develop an agreed test guideline for an alternative test species under circumstances where *Lemna* is not considered the most appropriate test species at Tier 1, that is, *Myriophyllum* species. Chair: Peter Dohmen.

Workgroup 3: Produce a database of existing methods using macrophytes based on the experience of participants and published literature. Collate information from experts via a questionnaire. Chair: Peter Ebke.

Workgroup 4: Develop guidance for the use of macrophyte data in species sensitivity distributions (SSDs). Chair: Stefania Loutseti.

Workgroup outputs are the results of actions agreed in plenary sessions of the AMRAP workshop. They provide a consensus view of those stakeholders participating in respective workgroups and have subsequently been reviewed by all participants.

The output from AMRAP is presented in the following sequence of chapters in a format that we hope will provide the reader with the appropriate amount of information necessary to inform their degree of interest.

1) Chapter 2 provides guidance on macrophyte risk assessment and summarizes the recommendations arising from the workshop with a brief rationale for each. It also explains the proposed decision scheme for aquatic macrophyte risk assessment.

2) Chapter 3 provides more detailed background information to elaborate the recommendations and guidance in Chapter 2.

3) Chapter 4 summarizes outputs from breakout groups and plenary discussions of three herbicide case studies that were used to explore approaches to characterizing risk in the context of aquatic macrophyte risk assessment.

4) Chapter 5 explains the background to and activities of the four workgroups set up at AMRAP whose work continues beyond the workshop.

5) Chapter 6 provides keynote papers that were used as an introduction to AMRAP, both to inform and to act as background information and thought provokers.

# 2 Guidance, Recommendations, and Proposed Decision Scheme for Additional Aquatic Macrophyte Tests

## 2.1 RATIONALE

There is clear scientific evidence to support the contention that aquatic macrophytes play a key role in the structure and functioning of aquatic ecosystems and, hence, must be considered within the risk assessment process for plant protection products (see Section 6.1). Under existing risk assessment procedures in the European Union (EU), the risk of herbicides to aquatic plants and algae is initially evaluated by calculating toxicity exposure ratios (TERs) between toxicity endpoints (EC50), derived from standard laboratory tests with 2 algae and one *Lemna* species, and predicted environmental concentrations (PECs). The resulting TER is compared with a trigger of 10, defined in Annex VI of 91/414/EEC. TER values that exceed this trigger indicate that the compound under evaluation can be considered to pose an acceptable risk to aquatic plants and algae, whereas TER values that fall below this trigger indicate a potential unacceptable risk and the need for a higher-tier risk assessment. However, there is concern that risk assessments based on *Lemna* endpoints may not be protective of other macrophyte species. Furthermore, there is a lack of guidance on the conduct and design of higher-tier studies focusing on aquatic macrophytes. Both issues were considered during the workshop.

A summary of the key points that were raised during the workshop and subsequent workgroup activities is presented in this chapter. These discussions have been used to formulate a series of recommendations and design a decision-making scheme to determine the need for additional tests with aquatic macrophytes. More details behind the decision scheme can be found in Chapter 3. While the focus is on herbicides and PGRs, the use of this scheme may be considered for assessing the risk of other chemicals, such as fungicides, that exhibit herbicidal activity.

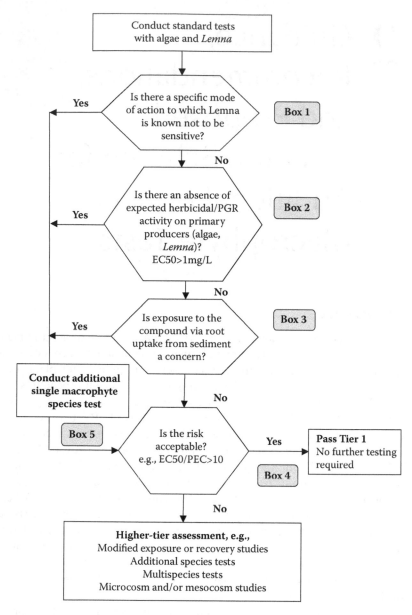

**FIGURE 2.1**    Proposed decision scheme for conducting additional aquatic macrophyte tests.

## 2.2   TIER 1: PROPOSED DECISION SCHEME FOR ADDITIONAL AQUATIC MACROPHYTE TESTS

The proposed decision scheme to determine the need for additional aquatic macrophyte tests in the risk assessment process is illustrated in Figure 2.1, where box numbers in the text refer to the decision points in the scheme. Under the existing risk assessment scheme for herbicides (including plant growth regulators [PGRs]), based on algal and *Lemna* toxicity data, a TER of 10 or above indicates an acceptable risk to aquatic macrophytes, and, therefore, further tests are not required. However, at Tier 1, false negatives are of regulatory concern; that is, the risk assessment using *Lemna* endpoints concludes that there is no unacceptable risk when in fact there is a risk.

There are three circumstances when this may be the case, and these are shown in Boxes 1, 2, and 3 in Figure 2.1:

**Box 1:** The herbicide or PGR has a specific toxic mode of action (MoA) (e.g., synthetic auxins or auxin inhibitors) to which *Lemna* is known not to be sensitive. (In such cases, *Lemna* is not a suitable test species.) If the answer to the question in Box 1 is yes, then a test with an alternative macrophyte species is required.

**Box 2:** There is an absence of expected herbicidal or plant growth regulatory activity in Tier 1 *Lemna* and algal tests.

The absence of expected activity in these tests may indicate that the standard test species lack the sensitivity of other primary producers. Under these circumstances, a risk assessment based on *Lemna* endpoints combined with an assessment factor may not be sufficiently protective of other macrophyte species, and a test with an alternative, sensitive macrophyte species is recommended.

If the answer to the question in Box 2 is yes, then a test with an additional macrophyte species is required.

**Box 3:** The chemical is known to partition to the sediment from the water column, and root uptake of the pesticide from the sediment is likely to be an important route of exposure.

*Lemna*, being a non–sediment-rooted macrophyte, may not respond in the same way to either positive or negative effects due to root uptake of the pesticide from the sediment.

If the answer to the question in Box 3 is yes, then a test with an additional macrophyte species is required.

Definitions of criteria that may trigger a test with an alternative macrophyte species have been elaborated by Workgroup 1 and are detailed in Section 5.1.

Recommendation 1: Conduct an additional test with a rooted macrophyte species where

- *Lemna* is known not to be sensitive to the test compound's MoA, or
- there is a lack of expected herbicidal or PGR activity in Tier 1 *Lemna* and algal tests, or
- exposure via sediment may be a critical factor in the risk assessment.

## 2.2.1 ALTERNATIVE SPECIES TEST AT TIER 1

Where the risk assessment triggers a test with an additional species (Box 5), there needs to be confidence in the ability of the test method to generate reliable and usable data. The workshop participants recognized that while several test methods have been developed to assess the effect of pesticides on rooted macrophytes, agreed test protocols using alternative macrophyte species are either not available or are under development. It is impossible to incorporate every desirable feature into a single test using a single species, and reasonable compromises have to be made while ensuring that the test is sufficiently robust to meet any shortcomings exhibited by a test using *Lemna*. The selected species should be a rooted macrophyte. Other factors to consider include availability from suppliers, ease of cultivation, and demonstration of measurable growth under controlled-environment conditions over the test duration. Several species, including *Myriophyllum*, *Glyceria*, and *Elodea* sp., have been used for research purposes. However, in discussions at the workshop there was clearly more experience with *Myriophyllum* than with other rooted macrophytes. It was decided, therefore, to support the activities of Workgroup 2 in the development and validation of a test method using *Myriophyllum*.

The current situation (2009) is that Workgroup 2 is arranging for a *Myriophyllum* test method to undergo ring-testing in a number of institutes with the anticipation that it will result in a validated method for testing plant protection products. The endpoint from this study would then be used for the assessment of risk to aquatic macrophytes. If this risk assessment (Box 4, Figure 2.1) indicates an acceptable level of risk (i.e., the answer to the question is yes), then further testing is not required. Eventually the test will be proposed as an OECD guideline.

Recommendation 2: Assess the effectiveness and reproducibility of an agreed test protocol for a rooted macrophyte (*Myriophyllum* sp.) via a ring test.

## 2.2.2 ECOTOXICOLOGICAL ENDPOINTS FOR A TIER 1 TEST

This critical and often contentious issue was debated vigorously by the workshop participants, who focused on three aspects:

- use of no-observed effects concentration (NOEC), ECx, or EC50 endpoints;
- choice of test duration relative to macrophyte growth rates; and
- choice of measurement parameters that are used to derive endpoints.

The use of a NOEC was considered to have well-recognized limitations even though its use is being promoted through revisions to Directive 91/414/EEC and the Water Framework Directive. Standard Tier 1 algal and *Lemna* tests are considered to provide a chronic assessment of toxicity because they cover several reproductive cycles in a short period (up to 7 days) and guidelines for both studies are designed to generate an EC50 endpoint. At present the risk assessment uses an assessment factor (TER) of 10, reflecting the fact that the test assesses chronic effects. AMRAP participants considered that, for statistical robustness, a lower ECx for example, an EC10 may be preferred to the use of an NOEC (Hanson et al. 2002). However, because existing test designs (based on fast-growing species, principally *Lemna*) are usually focused on the calculation of an EC50 they are not always suitable to determine a lower ECx value. Effects assessments with slower-growing submerged macrophytes are likely to require differently designed studies in order to generate appropriate measurement endpoints for use in risk assessment.

Because of the potential need to utilize data from several species within a species sensitivity distribution (SSD) analysis (Chapter 3), the view of the workshop participants was that evaluation of effects on growth should be based on assessments of biomass and shoot length because these generally provide consistency across time and species. Whatever measurement endpoint is chosen, the coefficient of variation should be low. Further guidance will be developed by Workgroup 4.

> Recommendation 3: Assess growth in additional aquatic macrophyte tests using biomass and shoot length measurements.

## 2.3  HIGHER-TIER RISK ASSESSMENT

For chemicals that fail the Tier 1 risk assessment, higher-tier assessments are required. Higher-tier refinements may take the form of mitigation of exposure through implementation of buffer zones or drift reduction techniques resulting in a reduction in PECs. Alternatively or additionally, higher-tier studies may be conducted to further evaluate the toxicity of the test compound to a wider range of macrophyte species and/or to evaluate effects under more realistic exposure conditions. The relative merits of different types of higher-tier studies were considered by workshop participants, both as a consequence of the experience of delegates and from the output of case studies that were evaluated during the workshop (Chapter 4).

When additional studies using macrophytes are being considered, a series of questions must be addressed in order to design a study that will answer the issues raised by the chemical and its use pattern. These questions are detailed in Section 3.1.1.2. In summary, the studies should be designed to provide endpoints that can be interpreted from an ecotoxicological and ecological perspective while also enabling regulatory assessment of the level of effect.

### 2.3.1  Exposure Considerations

The exposure element in any ecotoxicology study is an important consideration if such studies are to account for the types of pesticide exposure profiles generated in surface waters from the use of the chemical. The SETAC-sponsored workshop ELINK (Brock et al. in press) has developed guidance as to how ecotoxicological study design can better reflect typical (generalized) pesticide concentration profiles in surface waters. For rooted macrophytes, where growth rates and reproductive cycles are slower than the floating macrophyte *Lemna,* it is important that the interaction between exposure and effect is captured. In additional single-species tests, issues such as time to onset of effects should be properly addressed, as should the length of the exposure period. This use of the exposure profile needs to be geared to the species of concern. For risk assessment based on slower-growing aquatic macrophyte species, the use of a time-weighted average (TWA) concentration, as opposed to an initial PEC or a maximum PEC, may be appropriate. Circumstances in which the use of a TWA PEC may or may not be appropriate are discussed in ELINK Chapter 1, Section 1.4.2. (Brock et al., in press).

> Recommendation 4: Consider the exposure profile in relation to the species and effect under investigation, consider the length of the study required in relation to the expected exposure profile, and take into account the ecological context of the scenarios under scrutiny when higher-tier studies are designed using modified exposure regimes.

### 2.3.2 SELECTION OF RELEVANT SPECIES

Tier 1 risk assessments may indicate a need to evaluate effects on a range of macrophyte species in single-species, multispecies, microcosm, or mesocosm studies. The workshop participants considered that it was important to define those additional species that may be suitable for use in higher-tier tests, in terms of their taxonomy, growth form, availability, and growth rate (capturing any responses within a defined test period). For this purpose, Workgroup 3 was charged with compiling a list of macrophyte species through questionnaires requesting information on researchers' experiences with a wide range of macrophyte species. The list is to be used to aid species selection for higher-tier testing, including the generation of SSDs. However, there is a need for further work to develop reliable test methods for different species in addition to the *Myriophyllum* test protocol that is already under development (Section 5.2).

> Recommendation 5: Collate a list of aquatic macrophyte species to guide the selection of appropriate species for evaluation in single-species, multispecies, or microcosm and mesocosm tests.

### 2.3.3 SPECIES SENSITIVITY DISTRIBUTIONS

Species sensitivity distributions are potentially useful tools to determine the relative sensitivity of a range of species to a test substance and, in particular, as a means of comparing the sensitivity of the current Tier 1 macrophyte *Lemna* with that of other species. Workshop participants discussed the potential use of SSD analyses in risk assessment and concluded that there are areas of uncertainty associated specifically with the use of macrophyte data, particularly the selection of species and endpoints.

#### 2.3.3.1 Species Selection

Ideally, SSDs should be based on comparable endpoints generated from tests conducted under similar exposure scenarios and exposure durations, preferably using standardized protocols. However, due to the diversity of aquatic plant morphologies and differing test species requirements, this approach often is not practical. Instead, consideration should be given to the inclusion of species of concern based on results of lower-tier assessments, MoA, selectivity, and ecological relevance. Workshop participants concluded that species included in the SSD should be representative of different growth habits and taxonomic groups while also being ecologically relevant to the exposure scenarios addressed in the risk assessment. However, discussion arising during the case studies also indicated that for compounds that are known to be selective for a particular group of species, for example, submerged species, it may not prove possible to fit a single SSD across a more diverse range of species. Under these circumstances, it may be necessary to focus on a less diverse group of species for the SSD analysis (Van den Brink et al. 2006). Selection of species should not be based on geographic distribution, but on their relevance to the ecosystem of interest, also recognizing that relative species sensitivities may differ in different ecological scenarios.

### 2.3.3.2   Endpoint Selection

Growth rate endpoints, based on biomass or shoot length, are recommended because they potentially provide consistency across time and species. From a statistical viewpoint, it is preferable that all endpoints used in development of an SSD are based on common measurement parameters because each parameter may have a different distribution. An alternative approach is to use the lowest endpoint, no matter what measurement parameter it is based on.

Selection of endpoints should also consider the MoA of the test substance. For example, the effects of auxin-simulating herbicides may lead to distorted growth but not necessarily to a reduction in biomass. In these cases, measurement parameters other than biomass may be more applicable. Measurement parameters, from which endpoints are calculated, should preferably be sensitive and responsive in the range of tested concentrations such that SSDs avoid the use of greater-than values (i.e., no effect observed at the highest treatment concentration). However, it is recognized that obtaining clear and reproducible dose–response curves with slower-growing macrophytes is often difficult and that the endpoint may be greater than the highest concentration tested. However, workshop participants considered that future studies should try to build in test concentrations to avoid greater-than values unless poor solubility or lack of response at concentrations of >100 mg/L were evident. Additionally, the use of biochemical endpoints or biomarkers was not recommended due to difficulties in correlating results with tangible ecological effects, hence making their relevance uncertain.

In order to provide further guidance on the use of macrophyte endpoints in SSD analyses, Workgroup 4 has undertaken compilation of a database of macrophyte endpoints from several sources. To date, data representing more than 2000 endpoints for 54 compounds, predominantly herbicides, in 55 freshwater aquatic macrophyte species have been added to the database. For each endpoint, the database contains a record of several parameters, including the statistical endpoint, the growth measurement method, for example, shoot length (increase), shoot length (final), shoot numbers (final), and increase in dry weight. Workgroup 4 will conduct analyses with these data in order to provide further guidance on the selection of species and endpoints for use in SSD analyses (Section 5.4).

Recommendation 6: Collate and analyze data on single-species macrophyte toxicity in order to enable an assessment of the relative sensitivity of *Lemna* to that of other macrophytes.

Recommendation 7: Include a range of morphologically and taxonomically different macrophytes in SSDs, unless it is known that a specific macrophyte group is at risk, in which case the SSD should focus on them. Where feasible, endpoints should be based on a common measurement.

Recommendation 8: Disseminate AMRAP guidance concerning the construction and use of SSDs for aquatic macrophytes with the aim of reaching agreement on SSD criteria and outputs for use within the regulatory framework.

### 2.3.4 MULTISPECIES TESTS INCLUDING MICROCOSMS AND MESOCOSMS

There are many examples of multispecies macrophyte microcosm and mesocosm experiments in the literature, and the AMRAP-phenyl urea case study included examples of indoor and outdoor studies (see Chapter 4 and Appendix I). Because the issues to be addressed within a multispecies study may be complex, it is essential to design the study appropriately to address issues such as recovery assessment, inclusion of relevant species, adoption of appropriate assessment methods, and treatment (exposure) regime.

Assessment endpoints in mesocosm studies are commonly based on shoot length and/or final biomass. Periodic assessment of effects and recovery in mesocosm studies can be enabled through the use of bioassays, whereby potted plants, held at different depths to reflect their natural habit, are removed at intervals for assessment. Alternatively, large-scale ponds can be established into which enclosures are introduced or mesocosm or microcosm systems can be developed in a replicated test design.

Species should be representative of those found in ecosystems of concern that will also be amenable to assessment of their sensitivity under experimental conditions. For higher-tier approaches that aim to assess the sensitivity of different species in a more realistic exposure environment, then, for example, a potted plant multispecies test using appropriate sensitive species may be a preferred approach. However, for assessments that require examining ecosystem-level effects on natural communities associated with a specified agricultural scenario, then the use of mature, replicated enclosures or microcosms and mesocosms with naturally established macrophyte communities or introduced macrophyte species may be more appropriate. Alternatively, a combination of approaches may be feasible. The advantages and limitations of each study design are detailed in Chapter 3. The possible inclusion of *Lemna* in mesocosm-type studies was considered. However, the growth conditions suitable for the study of rooted macrophytes may not be optimal for the growth of

*Lemna*. While acknowledging that in some circumstances *Lemna* could be used for determining an endpoint in a mesocosm study, workshop participants agreed that separate *Lemna* bioassays or other higher-tier studies with *Lemna* would generally be more appropriate.

Recommendation 9: Ensure that mesocosm studies are appropriately designed to answer questions concerning either effects on sensitive specific species, for example, using a potted plant approach, or effects (direct and indirect) on natural (established) communities, or a combination of both approaches. Mesocosm studies addressing risks of herbicides or PGRs should contain a sufficient variety of morphological forms and taxonomic groups to enable an adequate assessment of risk.

Recommendation 10: Include *Lemna* in mesocosm studies where feasible and appropriate. Where conditions for the growth of submerged macrophytes are not optimal for *Lemna*, separate bioassays or other higher-tier experiments using *Lemna* may be used.

### 2.3.5 ECOLOGICAL CONTEXT

Regardless of the approaches used to assess the risk of chemicals to aquatic macrophytes, it is important that results can be extrapolated to natural ecosystems in both a spatial and temporal context. This task is highly complex because of the heterogeneity of agricultural landscapes and associated water bodies. To some extent, this task demands the linking of exposure potential and the likelihood of consequent effects. This link has been attempted in another SETAC Europe workshop ELINK (Brock et al., in press), where generalized exposure scenarios have been developed to inform ecotoxicology so that tests can be designed to better reflect the exposure profile predominating in a landscape.

In another SETAC Europe workshop (AMPERE), the relevance of microcosm and mesocosm studies and their role in regulatory risk assessment was debated. Published outputs are not available from this workshop because the rationale behind it was to provide a forum for open debate between key stakeholders concerning the value of such studies in aquatic risk assessment. While discussions primarily focused on criteria related to effects on aquatic invertebrate data (the most commonly generated mesocosm data sets), the same issues apply to aquatic macrophytes. The usefulness of such studies in the examination of direct and indirect effects of pesticides on populations, communities, and ecosystems is not in doubt, but their ability to reflect the uncertainties of real-world interactions and, importantly, the potential for recovery, requires further clarification. In particular, we need improved descriptions of aquatic landscapes of the kind outlined by Jeremy Biggs for the UK (Section 6.1). We also need to characterize species distributions in these landscapes, and critically, we need to generate confidence in our ability to assess whether the effects on sensitive species observed within a mesocosm study are realized in different ecological scenarios.

Recommendation 11: Develop tools for the temporal–spatial extrapolation of mesocosm data in order to gain a better understanding of the ability of mesocosms to reflect macrophyte responses in natural systems.

## 2.4   INFORMING DECISION-MAKING

Building regulatory confidence in methods for assessing the risk of pesticides to aquatic macrophytes is a major goal of this publication and the ongoing workgroups. It is also envisaged that short courses will be organized to aid communication, increase knowledge exchange, and build confidence. The AMRAP workshop demonstrated that there is already a wealth of knowledge within the scientific community concerning aquatic macrophytes, which will act as a springboard to continue the development of methods and assessment tools to assist decision-making. For this reason, the establishment of a SETAC Aquatic Macrophyte Advisory Group on aquatic macrophyte ecotoxicology was proposed by the workshop participants. For this purpose, the Aquatic Macrophyte Ecotoxicology Group (AMEG) has been formed and will act as a focal point for ongoing discussion and development of the science of risk assessment for aquatic macrophytes.

Council Directive 91/414/EEC is currently being revised, and the EU Guidance Document on Aquatic Ecotoxicology (EC 2002) will be revised over the next few years. The issues discussed in this workshop are pertinent to these revisions and should be considered accordingly.

Recommendation 12: Establish an aquatic macrophyte advisory group under the auspices of SETAC to continue the development of risk assessment tools and to steer education and training in aquatic macrophyte ecotoxicology.

# 3 Regulatory Issues Concerning Effects of Pesticides on Aquatic Macrophytes

This chapter builds upon the issues addressed in Chapter 2 and provides additional detail. It also defines the current state of knowledge concerning aquatic macrophytes in the context of pesticide risk assessment.

## 3.1 WHY MACROPHYTES ARE IMPORTANT IN REGULATORY AQUATIC RISK ASSESSMENT

Macrophytes are key species in ecosystems because they maintain biodiversity and fulfill important functional roles (see Chapter 1). They provide habitat, food, and spawning substrate, and macrophyte heterogeneity promotes aquatic biodiversity. Macrophytes also affect the chemical and physical properties of aquatic systems. Photosynthetic activity and other metabolic processes can change water chemistry by affecting the dynamics of dissolved gases (oxygen, carbon dioxide), hydrogen ions (pH), and nutrients (phosphate, nitrogen) and hence may influence purification and detoxification processes in aquatic systems. The growth of rooted macrophytes modifies flow conditions, increasing sedimentation rates, and stabilizes sediments. The distribution of aquatic macrophytes in any waterbody is also influenced by landscape, soil and water chemistry, climate, and local agronomic practices. In order to protect aquatic ecosystems in agricultural landscapes, and to sustain biodiversity and the ecological functions outlined above, it is vital that we understand the risk that plant protection products pose to aquatic macrophytes.

Under the Council Directive 91/414/EEC (EU 1997) and its revision, there is a requirement to protect macrophytes. The Directive states that a test on higher aquatic plants (i.e., *Lemna*) must be performed, especially for herbicides and plant growth regulators (PGRs). However, is this requirement sufficient? How is the regulatory need to consider the risk to aquatic macrophytes met when either the first-tier assessment fails and/or the current approach, based on *Lemna* and algae, is not suitable to assess the risk to macrophytes?

### 3.1.1 KEY REGULATORY ISSUES

#### 3.1.1.1 Tier 1 Assessments

*3.1.1.1.1 False Negatives*

False negatives (i.e., concluding that a substance does not cause unacceptable risk when it does) are the main regulatory concern for Tier 1 assessments. Uncertainty exists as to whether the current Tier 1 approach based upon toxicity values for *Lemna* species, plus an assessment factor, is sufficiently protective. Sources of uncertainty include the exposure route and the mode of action (MoA), the latter being broader than the toxic site of action because it also includes uptake by macrophytes and translocation and metabolic processes within plants.

Clear decision-making criteria are required to decide whether the exposure route and MoA are really of concern, which may require additional focused research. If concerns are justified, regulation should be adapted by, for example, including additional standard test species, which will generate the need to develop standardized methods for these additional species. The EU Aquatic Guidance Document (EC 2002) states the following:

> If there is evidence from efficacy data or data on terrestrial plants that the data for *Lemna* are not representative for other aquatic plant species (e.g., auxin simulators which can be more toxic to submerged plants than to *Lemna*; Belgers et al. 2007), additional data with other aquatic plant species may be required on a case-by-case basis. The test protocol for such studies should be discussed with the RMS or the competent authority because no internationally accepted guideline is available.
>
> At present, laboratory toxicity methods with aquatic macrophyte taxa other than *Lemna* are at an early stage of development, and will require further research before it is possible to develop a harmonized guideline. A protocol using *Myriophyllum* is being developed. However, notifiers are advised to discuss the study design with the Rapporteur Member State (RMS).

As summarized in Chapter 2, there are 3 circumstances in which an underestimate of risk may exist based on a false negative:

1) The chemical has a known MoA to which *Lemna* is not sensitive. *Lemna* may be insensitive to herbicides with certain modes of action such as synthetic auxins or auxin inhibitors. For these compounds, *Lemna* is considered an unsuitable representative for other aquatic macrophytes.
2) For an herbicide, there is an absence of expected toxicity to algae and *Lemna*. This might indicate that *Lemna* may not be representative of other macrophytes in terms of sensitivity to the herbicide.
3) The exposure route via the sediment is an important route for plant uptake. Adsorptive and persistent herbicides may potentially accumulate in sediments. Because *Lemna* is a non-sediment-rooted macrophyte, it may not respond to negative or positive effects of pesticides in sediment on aquatic macrophytes.

Participants at the workshop expressed concern that *Lemna*, being a non–sediment-rooted monocotyledon, may not be sensitive to modes of action unique to dicotyledonous (dicot) species and may not be as sensitive as fully submerged species that have a greater surface area exposed to the pesticide. In addition, because *Lemna* is not rooted in the sediment, negative or positive effects of pesticides taken up via the root system will not be evident. On the basis of limited existing data, differences between dicots and monocots seem to be either not relevant or less relevant within submerged aquatic macrophytes (Belgers et al. 2007; Arts et al. 2008). However, for emergent or floating species that may intercept spray drift, the differences in herbicide selectivity seen in terrestrial plants are more likely to be reflected in the responses of these groups of aquatic species.

### 3.1.1.1.2  Decision-Making Criteria

The need for decision-making criteria was identified in order to clarify the circumstances in which further data on other macrophytes are necessary. The workshop participants concluded that focused research is required to elaborate these criteria. If concerns that the current Tier 1 approach does not adequately address the risk to aquatic macrophytes are justified, then additional tools and guidance will be required. There is considerable experience with *Myriophyllum* sp. in terms of its use in assessment of effects of pesticides. Because *Myriophyllum* is a rooted macrophyte and a dicot species, it was considered to be an appropriate additional macrophyte test species for assessment of herbicidal activity because it may address both the sediment route of exposure and MoA issues. However, if *Myriophyllum* is considered to be a suitable additional Tier 1 test species, then a test guideline needs to be developed and accepted internationally.

### 3.1.1.1.3  Ecotoxicological Endpoints

An important question when considering any risk assessment procedure is what the appropriate ecotoxicological endpoint should be. Should the NOEC or an ECx (e.g., EC10 or EC50) be used as the regulatory endpoint? Current aquatic macrophyte risk assessment under Council Directive 91/414/EEC (EU 1997) uses the EC50 as the relevant endpoint. However, the Water Framework Directive uses the NOEC, and the revision of Annex II and III of Council Directive 91/414/EEC also proposes the use of NOEC for algae, *Lemna*, and other macrophytes because these tests are chronic studies for risk assessment. The use of a NOEC has several well-recognized limitations, a major one being its dependency on the specific test concentrations used in the experiment. For statistical robustness, determining an ECx may be more appropriate. The use of lower ECx values also has limitations, especially in cases where the lower end of the concentration–response curve is highly variable, and hence the uncertainty associated with a lower ECx may be high. In addition, current short-term toxicity tests are designed primarily to determine the EC50, and their design may be inappropriate for use in the determination of lower ECx values.

### 3.1.1.2  Higher-Tier Assessments

### 3.1.1.2.1  General Issues

If the first tier raises concerns, risk reduction measures may be considered in the form of exposure mitigation, either by using buffers or drift reduction measures or by refining the risk through higher-tier effects studies. Higher-tier effects studies may include modified exposure studies, additional species tests, and analysis to generate species sensitivity distributions (SSDs), multispecies tests, or microcosm and mesocosm studies. Clear guidance for higher-tier studies with aquatic macrophytes is not available, and some of the issues associated with the generation and interpretation of data from these studies are outlined in the case study evaluations in Chapter 4. If higher-tier aquatic macrophyte studies are conducted, then their design should be such that the information obtained from them can be adequately interpreted in relation to some or all of the following considerations:

- ability of the study to determine an effect level
- defining an acceptable level of effect
- inclusion of appropriate (realistic worst-case) exposure regimes
- reduction in uncertainty that the data generate in terms of deciding an assessment factor that could be used for spatial–temporal extrapolation
- in modified exposure studies, the selection of relevant species and endpoints
- selection of species and endpoints where data are intended to be used in a SSD approach
- defining the appropriate growth period and exposure profile
- determining whether an HC5 should be used
- in multispecies studies (including microcosm and mesocosm studies),
  - selection of test system, indoor or outdoor, size and complexity
  - species selection
  - appropriateness of a bioassay approach with potted species, compared with the use of naturally established replicated sediment and water enclosure systems or microcosms and mesocosms
  - study duration, exposure profile, and exposure time frame
  - choice of endpoints such as population, community, or ecosystem, or more than one level of biological organization
  - whether or not to measure or estimate recovery potential

### 3.1.1.2.2   *Species Sensitivity Distributions*

There is a need for methods for testing additional species, other than *Lemna* and *Myriophyllum* sp., in order to generate data for SSD analyses and to characterize the variation in macrophyte exposure and response between species. The development of new methods should draw on existing expertise with a range of species. The production of a list of additional test species providing information on taxonomy, growth form, availability, and potential test duration (based on growth) would help inform this process (Section 5.3).

If appropriate, species selection for the construction of SSDs should include species with different growth forms and taxonomy. Selection of species does not need to be based on geographical distribution, because SSD analysis shows that species from different geographical areas do not exhibit a systematic difference in sensitivity (Maltby et al. 2005). The number of species and method of calculation should follow established guidance (Maltby et al. 2005; Van den Brink et al. 2006). Several endpoints can be used to construct an SSD, if the endpoints are ecologically relevant. However, they should preferably include biomass, or growth rate estimates based on biomass or other morphological endpoints, like shoot length.

Two questions remain regarding the generation and analysis of SSDs for aquatic macrophytes:

1) Which species should be used to generate an SSD? Should both algal and macrophyte data, or subsets of them, be used to generate one SSD for primary producers? For which compounds, or groups of compounds, is this approach valid?

2) Should a common endpoint be used in the SSD? There is a need to evaluate the regulatory implications of constructing SSDs using either a common endpoint or the most sensitive endpoint for each species (see Chapter 4, Case Study AMRAP-Auxin).

The answers to these questions are also relevant to multispecies studies, and many of the recommendations provided for SSD analyses are equally applicable. However, differences between multispecies and single-species studies include the fact that larger plants and longer experimental periods may be used in multispecies studies, and hence, recovery may be investigated.

There was considerable debate among workshop participants on the inclusion of greater-than values in SSDs. Whenever possible, studies should be designed with exposure concentrations that minimize the generation of greater-than values. However, if greater-than or less-than values are valid endpoints, then it can be argued that they should not be ignored because doing so would bias the distribution of all data. It is also recognized that obtaining clear and reproducible dose–response curves with slower-growing macrophytes is often difficult and that the endpoint may be greater than the highest concentration tested. There are three options when considering greater-than values: exclude the data points, use data and ignore the greater-than sign, or include data in plots and in the hazard calculation but do not use them for curve-fitting. This issue was discussed in relation to case study AMRAP-SU (sulfonylurea, see Section 4.2.3), but agreement was not reached, and further consideration of the regulatory implications of including or excluding greater-than values is needed.

### 3.1.1.2.3   Recovery Potential

The relationship between the growth rate of a plant and its recovery potential was considered as part of Case Study AMRAP-SU. Workshop participants agreed that recovery potential following exposure to sublethal concentrations is correlated with growth rate, with slower growth generally indicating slower recovery. A consequence of this link between growth and recovery is that, because *Lemna* has a fast growth rate, recovery rates determined for *Lemna* may not be representative for slower-growing macrophyte species. The ecological effects of retarded growth and reduced biomass of a slower-growing macrophyte over a prolonged period need to be considered in the risk assessment.

*3.1.1.2.4    Multispecies Microcosm and Mesocosm Studies*

General guidance is available for the design and interpretation of microcosm and mesocosm studies (Campbell et al. 1999; Giddings et al. 2002; de Jong et al. 2008), although specific recommendations for higher-tier studies with aquatic macrophytes are not available. Microcosm and mesocosm studies can be used to evaluate community-level effects and species interactions. However, the number of macrophyte species in a mesocosm study may be limited, especially when population- and ecosystem-level effects are the focus of study. Therefore, tools for developing spatiotemporal extrapolation are necessary. An important question is: When *Lemna* is the most sensitive species in the SSD, should *Lemna*, or a similar species, be included in microcosm and mesocosm studies? The possible inclusion of *Lemna* in mesocosm macrophyte studies was considered. *Lemna* flourishes in eutrophic (nutrient-rich) surface waters, which also encourages the growth of algae, which in turn may lead to suppression of the growth of submerged macrophyte species. Submerged macrophytes and *Lemna* require different optimum abiotic growth conditions (Arts et al. 2001; van Liere et al. 2007). The optimum growth conditions for *Lemna* are characterized by high nutrient status, whereas optimal conditions for submerged macrophytes are characterized by moderate nutrient status with a nutrient-poor water layer over a nutrient-rich sediment (Arts et al. 2001; van Liere et al. 2007). If *Lemna* is to be studied with submerged macrophytes, then *Lemna trisulca* may be the most suitable species. Alternatively, separate *Lemna* bioassays or additional higher-tier studies with *Lemna* may be appropriate.

### 3.1.1.3    Communication and Knowledge Transfer

The refinement of existing methodologies and the development of novel techniques to address issues of regulatory concern are to be encouraged. However, in order for the regulatory community to develop confidence in the application of new risk assessment methods, every effort must be made to ensure clear communication and effective knowledge transfer. Novel techniques should be published and presented to the regulatory community and may form the basis for master classes, short courses, and workshops. Effective communication among all stakeholders is required throughout the development process in order to maximize the probability that novel techniques will be incorporated into the regulatory risk assessment framework.

### 3.1.2    KNOWLEDGE GAPS

The issues outlined in Section 3.1.1 above have identified a series of knowledge gaps that require further elaboration because they are important for risk assessment. Many of these are addressed in Section 3.2 and by the AMRAP workgroup reports in Chapter 5, and many will be the subjects of ongoing research into aquatic macrophyte risk assessment. The development of further knowledge may also help to underpin the decision-making scheme currently proposed in Figure 2.1.

- Clarification of which modes of action require testing with species other than *Lemna* (Workgroup 1).

- Information on the relative importance of sediment and water as exposure routes for rooted macrophytes and substances with different fate properties. Criteria are needed to determine when sediment exposure should be considered; these criteria may include $K_{OC}$, $K_{OW}$, and/or persistence (Workgroup 1).
- Development of a scientifically underpinned, standard protocol for an additional single-species macrophyte test for use in Tier 1 when *Lemna* is not appropriate (Workgroup 2).
- Guidance on the design of additional species tests (e.g., for SSD approach) (Workgroup 3).
- Collation of currently available information on the relative sensitivities of macrophytes to pesticides, specifically the relative sensitivity of *Lemna* species in relation to other aquatic macrophytes (Workgroup 4).
- Generation of comparative macrophyte SSD studies for substances that differ in their toxic MoA (Workgroup 4).
- Clarification on relevant endpoints (e.g., biomass, growth rate, root length).
- Identification of the advantages and limitations of multispecies tests with macrophytes and their relevance to natural ecosystems.
- Guidance on the design and interpretation of higher-tier macrophyte studies.
- Development of scientifically underpinned methods or tools for spatio-temporal extrapolation of mesocosm data, especially for the macrophyte component in those studies.
- Guidance on how to link exposure to and effects on aquatic macrophytes (e.g., exposure routes to consider: drift, run-off, drainage; effects of time-varying exposures; which PECs should be used in risk assessment).
- Guidance on the incorporation of recovery data into the risk assessment and extrapolation from recovery studies with *Lemna* to other species.
- Clarification on the use of assessment factors at higher tiers.
- Agreement on when an impact on macrophytes in the absence of pronounced indirect effects is acceptable, and the duration of these effects.

## 3.2  METHODOLOGIES: STRENGTHS AND WEAKNESSES

The expertise and experience of workshop delegates was explored in order to gain a better insight into test methods currently used or available for the assessment of the toxicity of pesticides to aquatic macrophytes. This information was used to identify the strengths and limitations of each approach, to define the gaps in knowledge and data requirements, and to consider the implications for risk assessment.

The specific questions addressed were these:

1) What single-species laboratory tests are currently available or are being developed, and what are their limitations?
2) What are the main differences between *Lemna* and other macrophytes in terms of life-history traits, recovery, experimental variability, and sensitivity?

3) What criteria should be considered when developing new test methods?
4) Which species and endpoints should be used in SSDs?
5) What is the representativeness and sensitivity of macrophyte species used in microcosm, mesocosm, and semi-field studies?

### 3.2.1  WHAT SINGLE-SPECIES LABORATORY TESTS ARE CURRENTLY AVAILABLE OR ARE BEING DEVELOPED?

The only protocol currently available for use when a test on an aquatic macrophyte is required for regulatory purposes is for *Lemna* sp. (OECD 2006c). However, as discussed in the previous chapters, *Lemna* may not be the most appropriate species when either sediment is the main exposure route or when *Lemna* is not sensitive to a specific mode of toxic action of the test substance. In these cases, other species (e.g., submerged macrophytes) might be more suitable than *Lemna*, because of their different morphology or sensitivity.

#### 3.2.1.1  Available Test Protocols

Test protocols for testing alternative macrophyte species are available or under development, but none have been validated by a ring-test. They include 3 protocols using *Myriophyllum* (Poovey and Getsinger 2005; Skogerboe et al. 2006; ASTM Guide E 1913-04, ASTM 2007) and one using *Glyceria* (Davies 2001; Davies et al. 2003). Several laboratories have used the above protocols, for different purposes. ASTM E1913-04 is a 14-day test that assesses the phytotoxicity of chemicals to *Myriophyllum sibiricum* grown in sterile liquid growth medium containing sucrose. Researchers at the German Umweltbundesamt (UBA) have modified the ASTM protocol and used it in non-sterile conditions, without the addition of sucrose, but without satisfactory results (Maletzki et al. 2008). Plant growth measured as increased shoot length could be observed, but there was a decrease in biomass. Thus, this amended protocol is not recommended. In any case, inclusion of sucrose in media for toxicity tests with aquatic plants is not recommended due to potential negative feedback on photosynthetic pathways and the increased potential for bacterial and algal growth even when axenic cultures are used.

Other researchers have included sediment in the test system. A 10-day sediment-contact test with *Myriophyllum aquaticum* is described by Feiler et al. (2004), and experiences are positive with *Myriophyllum* and other species grown for 7 and 10 days (time needed to have a doubling of biomass in controls) in an artificial sediment (OECD 2004a, 2004b).

The presence of sediment and/or nutrients in the test media for macrophytes often results in microbial and algal development in the media and on the macrophytes (see Cedergreen et al., Chapter 6). These organisms influence the pesticide exposure because they are involved in their degradation and compete with macrophytes for nutrients and influence their growth. Tests have been developed by separating sediment and water, in order to minimize algal and bacterial development in the test medium. Using this approach, macrophytes can acquire their nutrients from the root

**TABLE 3.1**

Macrophyte species used in laboratory studies and potential suitability for single-species toxicity tests. Preferred species, based on amenability, are shown in boldface type. Species that are available from commercial suppliers are identified with an asterisk. Import licenses may be required for certain species in some countries.

| Floating | | Submerged, Non-rooted |
|---|---|---|
| **Lemna\* Spirodela\*** | | *Lemna triscula* |
| **Azolla\*** | | **Ceratophyllum *sp.* \*** |
| *Salvinia* | | **Chara** |
| *(Riccia)* | | |

| Submerged, Rooted | | Emergent |
|---|---|---|
| **Egeria\*** | *Lagarosiphon\** | **Glyceria\*** |
| **Elodea\*** | **Heteranthera** | *Sparganium* |
| *Hydrilla\** | *zosterifolia\** | *Sagittaria sagittifolia* |
| *Callitriche* | *Hygrophila* | *Phragmites* |
| *Ranunculus* | *Najas* | |
| *Potamogeton crispus,* | *Vallisneria\** | |
| **Myriophyllum\*** | *Hottonia palustris* | |

medium (Barko and Smart 1981a), while algal and bacterial growth is reduced in the shoot medium, which is poor in nutrients.

Further unpublished test methods and protocols have been developed in research centers and by industry, covering a range of species. AMRAP workshop participants had experience with a range of macrophyte species, which could be used for single-species tests (Table 3.1). "Preferred species" (a species that is representative of a certain growth habit and for which there is some experience in its use in toxicity tests) that are readily available included surface-floating species (i.e., *Lemna, Spirodela* and *Azolla*), submerged non-rooted species (i.e., *Ceratophyllum, Chara*), submerged rooted species (i.e., *Egeria, Elodea, Myriophyllum, Heteranthera*), and rooted, emergent species (i.e., *Glyceria*). However, standardized or validated methods for many of these species in toxicity testing are not available. These issues are being followed up by Workgroup 3. The majority of species listed in Table 3.1 are available from commercial suppliers.

### 3.2.2 WHAT ARE THE MAIN DIFFERENCES BETWEEN *LEMNA* AND OTHER MACROPHYTES IN TERMS OF LIFE-HISTORY TRAITS, RECOVERY, EXPERIMENTAL VARIABILITY, AND SENSITIVITY?

The main differences between *Lemna* and other macrophytes are listed in Table 3.2.

**TABLE 3.2**

**Main differences between *Lemna* and other aquatic macrophyte species**

| Parameter | *Lemna* | Other Macrophytes |
|---|---|---|
| Life history | Floating plant: uptake of nutrients usually from water | Floating, submerged, or emergent |
| | Mainly vegetative reproduction | Vegetative or sexual reproduction |
| | Population growth | Individual growth |
| | r-strategist | More k-strategist (relative) |
| Recovery (of growth) | Fast from a few fronds | Variable |
| | Large dispersal potential | Dispersal depends on the season |
| | | Vegetative dispersal and seed bank in sediment |
| Experimental variability | Small in laboratory tests (5%–10%) | Larger: 20%–70% |
| Sensitivity | Medium to more sensitive under optimal condition for tested MoA[1], (PSII[2], fatty acid[3], ALS[4], microtubulin[5], PSI inhibitor[6]) | Unknown for many macrophyte species, and sensitivity may differ per species and per compound |

[1] Mode of action.
[2] PSII = Photosystem II-inhibitor.
[3] Fatty acid = fatty acids synthesis inhibitor.
[4] ALS = acetolactate synthase inhibitor.
[5] Microtubulin = compounds that interfere with microtubule assembly.
[6] PSI inhibitor = Photosystem I-inhibitor, superoxide free radical production.

### 3.2.3 WHAT CRITERIA SHOULD BE CONSIDERED WHEN NEW TEST METHODS ARE DEVELOPED?

There was agreement among the workshop participants that one test protocol for a particular growth type, for example, rooted species, could be expected to be applicable for related species, with only minor modifications according to species used in the test. The main barriers to the development of single-species tests were identified as algal contamination, ease of endpoint measurement, reproducibility and variability, and the availability of suitable test species.

New single-species test methods used in addition to, or in replacement of, the *Lemna* test should fulfill the following requirements:

- Test species available throughout the year
- Test species easy to culture or cultivate
- Uniformity of plant material
- Appropriate and easily measured endpoints
- Acceptable growth over defined test period
- Acceptable coefficients of variation (CVs)
- Ring-tested and validated test method

- Reproducibility of test (standardized water and sediment)
- Include verification of exposure concentrations
- Amenable to media renewal or pulsed-dose options if necessary

For Tier 1 studies, some participants proposed that a worst-case assessment necessitated exposure of the submerged foliage to the pesticide via the growth medium in the absence of sediment. However, the growth of submerged, rooted species may not be optimal in the absence of sediment-anchored roots. Further discussion of this approach is required, although it is clear that, for compounds where sediment exposure may be important, the test species should be rooted in sediment. Possible standard media included Algal Assay Procedure (AAP; USEPA 1971, 1996); M4 (OECD GL 201 (OECD 2006a)); ISO 8692 (ISO 2004), M4 macro, Smart and Barko medium (Smart and Barko 1985), and standard artificial sediment, that is, OECD *Chironomus* artificial sediment (OECD 2004a).

### 3.2.4   WHICH SPECIES AND ENDPOINTS SHOULD BE USED IN SSDS?

Ideally, SSDs should be based on comparable endpoints generated from tests conducted under similar exposure scenarios and exposure durations, preferably using standardized protocols. Available data from aquatic plants, whether macrophytes or algae, should be evaluated to establish their relevance and whether endpoints belong to the same distribution. If algae and macrophytes clearly have different sensitivity distributions, then they should be evaluated separately. Consideration should be given to the inclusion of species of concern based on the results of lower-tier assessments, compound MoA, selectivity, ecological relevance, or other information.

Assuming that the endpoints can be described by the same distribution, species should represent different growth habits (submerged, emergent, floating, rooted, and non-rooted) and taxonomic groupings (monocotyledons and dicotyledons) and should cover as many genera as possible. Where a specific group of macrophytes, such as submerged species, is more sensitive to a compound than other taxa such that all species do not belong to the same SSD, then a number of SSD analyses may need to be conducted. Van den Brink et al. (2006) indicated that combining sensitive and non-sensitive taxa in the same SSD leads to a mismatch and lack of fit. Selection of species should not be based on geographic distribution but on their relevance to the ecosystem of interest. The number of species used to construct the SSD and method of calculation of HCx values should follow established guidance for SSDs (Maltby et al., 2005; Van den Brink et al. 2006).

Endpoints used in an SSD should be ecologically relevant but will, of course, be based on available data. Growth rate, based on biomass or shoot length, is the recommended endpoint because it potentially provides consistency across time and species. Due to the variation in macrophyte morphology, biomass is often the only common relevant measurement across species. Endpoints should be reliable with acceptable variability (to be defined by Workgroup 4). Greater variability is observed in root weight than shoot weight, and lowest variability is usually obtained for biomass or growth (Hanson et al. 2003; Knauer et al. 2006; Arts et al. 2008).

Selection of endpoints should consider the MoA of the test substance, along with time to effect for different endpoints. For example, the effects of auxin-simulating herbicides may lead to distorted growth but not necessarily to a reduction in biomass. In these cases, measurement parameters other than biomass may be more applicable.

From a statistical viewpoint, it is preferred that all endpoints used in development of an SSD are based on common measurement parameter (e.g., total shoot length) because each parameter has its own distribution. This view is countered by practical considerations, where the lowest endpoints, regardless of measurement parameter, are often used to generate the SSD.

Measurement parameters, from which endpoints are calculated, should preferably be sensitive and responsive in the range of tested concentrations such that SSDs (where possible) do not include greater-than values (see Section 2.3.3.2). Generally, biomarker endpoints should not be used for risk assessment due to difficulties in establishing their ecological relevance. These endpoints are considered more relevant for mechanistic studies or hazard assessment.

### 3.2.5  WHAT IS THE REPRESENTATIVENESS AND SENSITIVITY OF MACROPHYTE SPECIES USED IN MICROCOSM, MESOCOSM, AND SEMI-FIELD STUDIES?

There have been several good examples of testing macrophytes in predominantly outdoor microcosms and mesocosms. In all cases, it is important to be able to follow effects and recovery over time. With regard to recovery assessment, if the growth of macrophytes is not limited in controls, then recovery cannot be assessed based on final biomass at the end of study because the biomass of the treated plants will never catch up with the control. In this case, it may be sufficient to demonstrate that the growth rate of the treated plants has recovered. Carry-over effects into the next season are not commonly assessed because many species enter senescence during the autumn months. This aspect is further complicated by the fact that some water bodies are dredged at the end of the growing season (e.g., Dutch ditches).

In general, two basic approaches have emerged for testing macrophyte responses in microcosm and mesocosm studies:

1) Potted plants in microcosms and mesocosms, so-called "multispecies tests." The macrophytes are planted in pots, either separately or as mixed species communities in individual pots. Typically up to 10 species can be tested in this method, depending on the size of the microcosms or mesocosms, and each treated mesocosm is used as a replicate. Pots can be exposed at different depths dependent on species needs; for example, *Lemna* can be investigated using ring enclosures, and larger floating species (e.g., *Ceratophyllum*) can be kept in cages.

2) Macrophytes grown in natural sediment enclosures or microcosms and mesocosms. Several species of plants are grown naturally in larger ponds introduced via the sediment or via planting of shoots or introduction of diaspores. Multiple enclosures or microcosms or mesocosms serve as replicated test units. It is also possible to introduce floating species in ringed

enclosures or cages (for free-floating species). Typically, this design allows a number of species to be analyzed quantitatively. However, because often only a few macrophytes dominate natural plant communities, the number of species to be analyzed quantitatively and statistically in such an experimental approach is limited. Macrophyte species may be harvested at the end of the experiment to quantify effects on biomass. The advantages and limitations of both approaches are given in Table 3.3.

## TABLE 3.3
## Advantages and limitations of assessing phytotoxicity in microcosms and mesocosms using potted plants or plants rooted in natural sediment

| Approach | Advantages | Limitations |
|---|---|---|
| Potted plants | Multiple species in individual pots can be used to assess species interactions. Intermediate time measurements are possible by either destructive or non-destructive sampling (e.g., length, wet weight) Statistical variability is lower compared to naturally grown populations, and a higher number of samples (number of individual pots) is possible. The relevant test species can be selected dependent on sensitivity and/or maximizing taxonomic diversity. Direct effects can be measured and recovery of individual plants can be followed. | Species interactions (competition) are not assessed optimally, which might affect sensitivity. For multispecies pots, it may be difficult to select species that are able to grow together. It might be difficult to manage the system (e.g., nutrient levels) to avoid over-growing of macrophytes by algae (e.g., filamentous). |
| Plants in sediment | This approach allows natural communities to be assessed including competition or indirect effects. This approach can be supplemented by use of potted plants (bioassays) within enclosures to assess direct effects. | Intermediate time measurements are possible only by nondestructive sampling (e.g., monitoring of covered surface area). Destructive sampling is possible only at the end of the study so only one biomass measurement is possible. Variability may be higher than with the potted plant approach. Indirect effects might mask direct effects. |

Species used in microcosm and mesocosm studies should be selected based on the specific question that needs to be addressed and on their sensitivity and/or representativeness for natural ecosystems. For example, if the aim is to refine the lower-tier risk assessment through determining the sensitivity of different species with a focus on the most sensitive species, then the single-species potted plant approach may be the most appropriate method. It should be noted that not all of the species that can be studied in single-species laboratory studies can be kept under outdoor conditions (e.g., tropical species). Alternatively, if the focus of the assessment is to investigate effects (in particular, indirect or competitive effects) on natural communities within a given waterbody type (e.g., pond, ditch) and/or in a particular geographical region, then the natural sediment enclosure approach may be more appropriate as well as a mesocosm study including natural or introduced macrophyte populations.

# 4 Characterizing and Assessing Risk Using Case Studies

## 4.1 INTRODUCTION

As with any assessment of the risks of pesticides to an environmental compartment, it is essential that adequate data of the appropriate type are available in order to inform decision-making. It is, however, often the case that an ecotoxicological study will not provide all the information necessary to answer a risk assessment issue because either it has been designed in response to a different question or it completes only one piece of the jigsaw puzzle. The questions below (which can be applied to the case studies in this chapter) are general questions that can be asked when a risk assessment is conducted for aquatic macrophytes, in order to assess whether sufficient and appropriate information is available.

- Are the exposure scenarios relevant to the effects assessments?
- Can endpoints from laboratory and multispecies or mesocosm studies be compared in terms of sensitivity?
- Are exposure data appropriate or adequate for risk assessment?
- Have the appropriate studies been done at the right stage of risk assessment?
- Has sufficient account been taken of the necessity for assessment of recovery?
- Have the appropriate macrophyte species been studied?

Risk assessment case studies are useful tools in that they focus attention on specific risks or properties of a pesticide together with solutions that are considered to either resolve the issue or highlight continued uncertainty. The AMRAP case studies focused on 3 herbicides from different chemical classes (shown in detail in Appendix 1) that raised concerns in the current Tier 1 assessment and thus required further investigation:

1) AMRAP-Auxin has a mode of action (MoA) to which *Lemna* is known to lack sensitivity, and thus additional macrophyte species testing was warranted.
2) AMRAP-Phenylurea has a high toxicity to *Lemna*, and the TER triggered higher-tier assessments. It is also systemic, and the possibility of uptake via the sediment could not be answered by the use of *Lemna*.
3) AMRAP-SU (sulfonylurea) shows high toxicity to algae, and *Lemna* and the Tier 1 TER triggered a higher-tier assessment.

The three case studies and risk assessment issues arising from them are presented in Sections 4.2 through 4.4. These comprise an overview of the data and critical points, together with the output from both breakout groups and plenary discussions at the workshop.

## 4.2   CASE STUDY EVALUATIONS

### 4.2.1   AMRAP-Auxin

#### 4.2.1.1   Introduction

AMRAP-Auxin is based on an auxin MoA herbicide that is used to control dicotyledonous weeds in cereal crops. Data for this case study are presented in Appendix 1 and are summarized here. The compound has high water solubility (24 mg/L at 24 °C, pH7) and a low $K_{OC}$ of 60. It is not persistent in soil or water–sediment systems with half-lives of 13 and 31 days, respectively. In accordance with recommendations of the Aquatic Guidance Document (EC 2002), the Tier 1 data package includes standard tests with algal and *Lemna* species, as well as data for the submerged, rooted species *Myriophyllum spicatum*. Tier 1 test data indicate that algae are relatively insensitive to this herbicide (EC50 of 41 mg ai/L), while *Lemna gibba* and *Myriophyllum spicatum* are significantly more sensitive with EC50 values of 0.58 and 0.0125 mg ai/L, respectively. Tier 1 TER values based on maximum initial PEC values exceed the Annex VI (Directive 91/414/EEC; EU 1997) trigger of 10 for algae and *Lemna* species, whereas the TER value for *Myriophyllum spicatum*, based on the most sensitive EC50, falls below 10. Therefore, the potential risk of this herbicide to aquatic plants was evaluated further by consideration of the available higher-tier data.

#### 4.2.1.2   Higher-Tier Data

In addition to the *Myriophyllum spicatum* laboratory study that was considered in the Tier 1 risk assessment, two further laboratory studies were conducted with submerged macrophyte species. In the first of these studies, *M. aquaticum* was evaluated in a test system containing an artificial rooting substrate. In a further study, 9 submerged species, including *M. spicatum*, were tested in the absence of sediment. However, due to lack of growth in *M. spicatum* control cultures, endpoints were generated for only the 8 remaining species. Endpoints were based on assessments of shoot and root dry weight or length and root number. Endpoints from Tier 1 and higher-tier studies were used to generate SSDs for each assessment parameter. All available endpoints were included, although EC50 values lying outside the exposure range of the test (i.e., greater-than values) were omitted. The resulting median HC5 values ranged between 18.6 and 75.5 µg ai/L.

In addition, an outdoor microcosm study was conducted to evaluate the effects of the test substance on the growth of submerged *Myriophyllum* and *Potamogeton* species. Outdoor enclosures were filled with a layer of natural sediment overlaid with natural pond water. Young plants with roots were collected from natural ponds and transplanted into plastic pots containing natural sediment, which were placed on the sediment surface in each enclosure. The study incorporated 3 replicate enclosures per treatment, each containing 12 individually potted plants per species. Solutions of the test substance were mixed into the water column to give nominal concentrations of 0.01 and 0.1 mg ai/L. Assessments of plant fresh weight (shoot and root) and the number of plants exhibiting symptoms of toxicity were made 30 and 60 days after treatment. Ignoring hormetic effects that were apparent in *Potamogeton* on day 60, the no observed ecologically adverse effect concentration (NOEAEC) for both species was 0.01 mg ai/L. Significant stimulation of plant fresh weight was apparent in *Potamogeton* exposed to 0.01 mg ai/L on day 60.

The following data and its use were identified for comment:

1) results of laboratory studies and their use in SSD assessments,
2) mesocosm data and its use in risk assessment, and
3) the regulatory acceptable concentration (RAC) for use in risk assessment.

### 4.2.1.2.1 Results of Laboratory Studies and Their Use in Species Sensitivity Distribution Assessments

Laboratory studies were scrutinized in terms of the test methods used, the species chosen, and whether or not the selection of species adequately represented the range of macrophytes likely to be exposed in reality.

Four issues arose from this evaluation relating to the adequacy of laboratory data in relation to its use in risk assessment:

1) range of species in the SSD,
2) realism of the methodology used in the laboratory tests,
3) variety of assessment parameters measured in the tests, and
4) number of species in the SSD.

It was concluded that the range of species tested should be selected to reflect a wide range of morphological forms and taxonomic groups rather than a specific assemblage of macrophytes likely to be exposed in reality. In this case study, the SSD was based on the results of 3 laboratory studies in which several species from the same genus were tested. For example, the SSD contained data for 2 *Myriophyllum*, 2 *Potamogeton*, and 3 *Ranunculus* species. It was recommended that, rather than focusing on a limited number of genera, a wider range of species should be selected in order to better represent the range of macrophytes that may be exposed. Nevertheless, the tested species were acknowledged to represent monocotyledon and dicotyledon species as well as rooted and non-rooted species. It was also considered that the methodology used in the laboratory studies could be improved to better reflect more realistic exposure conditions. In this case, the SSD was based on results for 8 species, including rooted species, which were tested in a water-only system. From an environmental fate perspective, the method was considered to represent a worst-case exposure because the absence of sediment from the system would effectively maximize the availability of the test substance for foliar uptake from the water column. However, from a biological perspective, the absence of sediment may delay root formation, leading to suboptimal macrophyte growth, and hence, overestimate herbicide activity. Overall, it was concluded that the method produced data that were considered to provide a conservative assessment of toxicity. This conclusion was considered acceptable in this case because, despite the absence of realism in the form of sediment, plant growth was shown to be adequate.

For a number of species, conclusive EC50 values could not be established for every assessment parameter due to the absence of a dose–response relationship over the concentration range that was tested or the lack of applicability of some assessment parameters to some species. For example, calculation of root endpoints for *Lemna trisulca* was not possible due to practical difficulties in performing root assessments for this species. Consequently, by constructing SSDs based on the same endpoint for each species, the number of species in an SSD is limited to a maximum of seven. Recommendations from HARAP (Campbell et al. 1999) were that at least 8 data points were required for an SSD. If fewer than 8 data points are available, the fit of the SSD and the confidence limits around the HC5 should be considered in order to determine whether the lower limit of the HC5 could be used for risk assessment purposes or whether the application of a safety factor on the median HC5 is necessary. An alternative proposal was that an SSD could be compiled using the most sensitive endpoint for each species. This approach would produce an SSD based on endpoints for 9 species. Workgroup participants agreed that further work was required to evaluate the validity of this approach (see Section 5.4).

Overall, it was agreed that direct comparison of species sensitivity based on the laboratory data was complicated by the lack of common endpoints for all species and the use of different test methods (i.e., without and with sediment). In particular, the endpoint that was derived for the majority of species was based on shoot dry weight. However, shoot length was generally the most sensitive parameter but was not measured in all species. For example, assessments of shoot length are not applicable in *Lemna* species. Consequently, the group recommended that, where possible, tests with different species should aim to generate common endpoints under common test conditions in order to reduce uncertainty in the resulting SSD and risk assessment.

### 4.2.1.2.2  Mesocosm Data and Its Use in Risk Assessment

The objective here was to consider how the mesocosm data should be incorporated into the risk assessment and if the mesocosm data supported the conclusions of the SSD and/or added value to the risk assessment.

Because the mesocosm study incorporated only 2 species and 2 test concentrations over a 60-day period, the group concluded that the study had been designed to validate responses in *Myriophyllum*, the most sensitive species in the laboratory, and the less sensitive monocot species *Potamogeton* under more realistic and prolonged exposure conditions.

Several questions were raised because information considered relevant to interpretation of the results and relevant to risk assessment was missing. For example, while the effects on the growth of *Myriophyllum* appeared consistent with the EC50 values generated from the laboratory, the lack of consistency between the parameters measured in the laboratory and the outdoor study prevented a conclusive comparison of results. Similarly, comparisons between laboratory and mesocosm data for *Potamogeton* appeared to show that it is more sensitive under field conditions than in the laboratory. While measurements made on day 30 of the field study and on day 28 of the laboratory study were in agreement, further effects seen on day 60 of the field study were not predictable from the laboratory data. Therefore, it was concluded that while data from the mesocosm study were generally consistent with the laboratory-based data for *Myriophyllum*, results from the mesocosm study raised additional issues for *Potamogeton* species. The reasons for the apparent discrepancy between laboratory and field data for this species were not immediately apparent but may have resulted from the measurement of different parameters, the absence or presence of sediment, and/or differences not realized in the shorter-duration laboratory study compared with that of the mesocosm study.

For these reasons, the mesocosm study alone was not considered appropriate for deriving the final regulatory endpoint for *Potamogeton*. However, it was considered to provide supporting evidence for the responses seen in *Myriophyllum* in the laboratory studies.

### 4.2.1.2.3  *Regulatory Acceptable Concentration for Use in Risk Assessment*
The key issue in this case was to consider whether or not a Tier 1 risk assessment based solely on *Lemna* data would be protective of other macrophytes.

Based on the Tier 1 *Lemna* EC50 of 0.58 mg ai/L and the Annex VI (Directive 91/414/EEC; EU 1997) trigger of 10, herbicide concentrations of 0.058 mg ai/L would be considered acceptable and would not pose an unacceptable risk to aquatic plants. However, tests with 10 additional species revealed significant effects on 2 *Myriophyllum* species at concentrations below 58 µg/L. Significant effects were also apparent in the 2 species tested in the mesocosm study at 100 µg/L.

SSD analyses indicated that the HC5 based on shoot dry weight (0.052 mg ai/L) was similar to the acceptable concentration of 0.058 mg a/L based on the Tier 1 *Lemna* endpoint. The HC5 of 0.0186 mg ai/L based on the most sensitive endpoint of shoot length was approximately 3-fold lower than these values. However, *Myriophyllum* was approximately 10-fold more sensitive than the next most sensitive species and >100-fold more sensitive than the least sensitive species. Consequently, the shape of the curve at its tail is rather flat, and accordingly the HC5 has a very low lower limit HC5 of 1 µg/L. This observation, combined with the fact that the SSD was based on data for only 7 species, led some participants to conclude that 0.001 mg ai/L (LL HC5) should be used for risk assessment. Meanwhile, others believed that the median HC5 (0.018 mg ai/L) was sufficiently protective, especially because this median value was the lowest HC5, derived after consideration of all endpoints, and that only one species was affected at such concentrations. These participants also cited evidence from other evaluations suggesting that the median HC5 from SSDs is generally protective for single applications of non-persistent compounds.

Consequently, there was no agreement on a final RAC, but it was concluded that, in this case, the Tier 1 risk assessment based solely on *Lemna* data would not be protective of other macrophytes.

### 4.2.2 AMRAP-Phenylurea

#### 4.2.2.1 Introduction

AMRAP-phenylurea is a substituted phenylurea herbicide, active against broadleaf weeds and grasses. Data for this case study are presented in Appendix 1 and summarized here. This herbicide acts by absorption through roots and foliage, and it is systemic. It has a water solubility of 68.3 mg/L, a log $K_{OW}$ of 3, and a moderate $K_{OC}$ of 450. It partitions between surface water and sediment and is moderately persistent, with a DT50 in overlying water of 48 and 220 days depending upon sediment type. Tier 1 laboratory studies were carried out with *Pseudokirchneriella subcapitata, Chlorella vulgaris,* and *Lemna minor.* The toxicity of the herbicide to *C. vulgaris* and *L. minor* was similar with a 7-day EC50 of 7 µg/L. The risk to algae and aquatic plants was evaluated by calculating the TERs based on the 7-day time-weighted-average surface water PEC value generated from FOCUS SW Step 3. The TER for *Lemna minor* was 2.7, which indicated the necessity for higher-tier assessment.

#### 4.2.2.2 Higher-Tier Data

The higher-tier assessment for aquatic macrophytes comprised 3 studies, all focusing on potential effects on rooted macrophytes.

1) Study 1 was conducted with *Myriophyllum spicatum* and *Potamogeton perfoliatus.* Glass aquaria (600 L capacity) containing a layer of natural sediment overlaid with 50-cm–deep natural pond water were planted with 10 shoots of each species. After 7 weeks, microcosms were treated with a single application of the test substance and maintained for a further 5 weeks. Assessments of shoot biomass were made at test initiation and at the end. The respective biomass EC50s were 137 µg/L and 25 µg/L. The EC50 of *P. perfoliatus* was one-tenth that for *Lemna minor.*

2) In Study 2, effects of the herbicide on aquatic plants and algae were assessed in microcosms and mesocosms comprising glass aquaria (600 L), containing a layer of natural sediment overlaid with 50-cm depth of water. Plankton, macro-invertebrates, and *Elodea nuttallii* were added and acclimatized for 3 months prior to treatment with the test substance. Systems were treated twice weekly with the herbicide for 4 weeks, followed by a 7-week non-treatment phase. *Elodea* shoots were harvested after 11 weeks for assessment of fresh and dry weight. A separate *E. nuttallii* bioassay was conducted within the mesocosms using caged plant shoots. The NOEC for *E. nuttallii* from the main study was 15 µg/L and that from the bioassay was 5 µg/L.

3) Study 3 was a replicated outdoor ditch mesocosm study. Ditches were macrophyte dominated and were treated once every 4 weeks with a total of 3 herbicide applications with concentrations up to 50 µg/L. Macrophyte species composition and abundance were monitored at designated intervals. Of the 12 macrophyte species present, the dominant species were *Sagittaria sagittifolia, Myriophyllum spicatum,* and *Elodea nuttallii. Ranunculus, Potamogeton,* and *Polygonum* species were also abundant in

some mesocosms. *S. sagittifolia* and *M. spicatum* increased in abundance during the first 2 treatment periods, after which time *S. sagittifolia* showed signs of senescence in all mesocosms. Both *M. spicatum* and *E. nuttallii* dominated until the end of the season. No relationship between the total number of macrophyte species and herbicide treatment was evident, nor was there a significant difference in mean cover of macrophytes in any treatment compared with controls. There was a nonsignificant reduction in biomass at 50 µg ai/L after the second application.

The following points were identified for comment:

1) Whether the partitioning of the test substance between water and sediment should be considered in the design of higher-tier studies.
2) Appropriateness of species selection and methodology used in higher-tier studies.
3) Use of photosynthetic measurement as risk assessment endpoints.
4) Mesocosm data and its use in risk assessment.

In order to address these points, the relative merits of each of the 3 higher-tier studies were considered.

### 4.2.2.2.1 Study 1

The information on the test methodology of the study was poor, especially with respect to exposure and competition issues. The participants felt that a comparison of intrinsic species sensitivities would have been better addressed by individual exposure tests because a 7-week equilibration period should result in high plant densities and competition. The influence of competition on effect levels needs to be considered when comparing data on species generated under different experimental conditions. As an overall assessment, the assumption of risk cannot be negated by this study because *Lemna* in the Tier 1 study seems to be of highest risk, followed by *Potamogeton*, with the TERs below 10 for both species.

### 4.2.2.2.2 Study 2

In this indoor mesocosm study, apart from algae plankton and invertebrates, *Elodea nuttallii* was the only macrophyte studied. This investigation was supported by a bioassay on *E. nuttallii*, carried out in the water column for the first 3 weeks of the study. Because the test substance was applied twice for 4 weeks, followed by 7 weeks of non-treatment, the study was able to demonstrate effects following exposure and recovery after a worst-case exposure scenario.

*Elodea nuttallii* shoots exposed in the bioassay were about 3 times more sensitive than *Elodea* grown in the sediment. The workgroup considered that the difference in effect level was most probably due to the different exposure routes. Also, it was not possible to say whether reduced intra-specific competition in the bioassays contributed to higher growth and sensitivity. Plenary discussions on the influence of the study design of bioassays on macrophyte responses showed that considerable variation in response to herbicides can be demonstrated using shoots suspended in medium alone, compared with macrophytes rooted in a sand or sediment layer.

The data were regarded as consistent with the results of Study 1 but did not clarify the recovery of the most sensitive representative species or genera and specifically did not include *Potamogeton,* which was shown to be relatively sensitive in Study 1. Thus, the risk for these 2 representative species cannot be fully negated. The question as to whether the TER of 2.7 for *Lemna* EC50, and the low NOEC for *Potamogeton* are acceptable remains open.

### 4.2.2.2.3    Study 3

A large-ditch mesocosm study with 3 applications of the herbicide (4-week intervals), each followed by moderate flushing after 7 days, investigated 12 macrophyte species. No effect was observed up to the highest concentration (50 µg/L) on either macrophyte biomass or the composition of the 3 dominant species. The exposure scenario was regarded as realistic worst case for streams and ditches.

There was some criticism concerning the macrophyte composition because there were statistical differences in distributions before exposure began. While this observation may not have influenced the study in any specific way, the view of the group was that replicated systems should be similar prior to treatment or consequent effects may be difficult to assess. For the regulatory sensitive endpoints, the study was regarded as being of limited value because it was not appropriately representative. *Lemna* was not present in the ditches, and *Potamogeton* grew only in some ditches and was not specifically assessed. Interestingly, the NOEC for green algae was the same as in the laboratory study (5 µg/L), and the NOEAEC after recovery was 10 times higher (50 µg/L). This underpins the comparability and consistency of the exposure–effect relationships, but leaves open the question of recovery of floating macrophytes and *Potamogeton.* It was the participants' view that the concentration range was too low to demonstrate a comparability of measurement methods and sensitivities between this study and Study 1.

#### 4.2.2.2.3.1    Consider the Partitioning of the Test Substance Between Water and Sediment in the Design of Higher-Tier Studies    The participants considered that the DT50 in water, compared with that in the whole system, indicated that the herbicide did not substantially partition to the sediment but remained in the water column, meaning that exposure in the water phase was the main concern.

#### 4.2.2.2.3.2    Species Selection and Methodology Used in Higher-Tier Studies    The key issues were that *Lemna* was not included in any higher-tier study and that *Potamogeton,* found to be the most sensitive rooted macrophyte in Study 1, was not subsequently investigated.

#### 4.2.2.2.3.3    Use of Photosynthetic Measurements in Risk Assessment    The participants considered that photosynthetic parameters (evolution of oxygen) were probably of less value than biomass as a measure of effect. If this were to be employed as an endpoint in mesocosm studies, it would require setting up specific assays with specific species to trap evolved oxygen.

*4.2.2.2.3.4 Mesocosm Data and Its Use in Risk Assessment* The mesocosm (ditch) study was evaluated with particular reference to species composition and whether or not the data added value to the risk assessment. The key issues were that *Lemna* was not included in the study and that *Potamogeton* was not specifically assessed, leading to concern that appropriate species data for risk assessment had not been generated. However, the effects on other species seemed to indicate no major surprises in toxicological response. It was therefore concluded that, while the mesocosm data added some confidence in reducing uncertainty, there remained a level of concern that had not been fully addressed.

Overall, this case study demonstrates the need to think carefully about study design and species composition in relation to Tier 1 effect data. In this example, there was no clear path of investigation in the higher-tier assessments, and perhaps more robust independent assays using *Lemna* and *Potamogeton* would have better addressed the risk.

## 4.2.3 AMRAP-SU

### 4.2.3.1 Introduction

AMRAP-SU is a sulfonylurea herbicide that is used for the control of grass and broad-leaf weeds in cereals. Data for this case study are presented in Appendix I and are summarized here. The compound has high water solubility (480 mg/L at 20 °C, pH7), low $K_{OC}$ of 43, and soil and water–sediment half-lives of 24 and approximately 40 days, respectively. The herbicide is applied once a year, either in the spring or autumn at BBCH 12-25. Results of FOCUS SW Step 3 modeling indicate that maximum initial surface water concentrations will occur following applications in autumn-sown crops. Maximum initial concentrations of 1.83 and 1.15 µg/L arose from the D2 ditch and stream scenarios, respectively. The corresponding 7-day time-weighted average concentrations were approximately half the initial concentration (0.89 and 0.46 µg/L, respectively). Tier 1 toxicity data indicate that algae are relatively insensitive to this herbicide (minimum EC50 of 65 mg ai/L), while *Lemna gibba* is significantly more sensitive with EC50 values of 1.5 to 2.1 µg ai/L. Tier 1 TER values based on maximum initial PEC values exceed the Annex VI (Directive 91/414/EEC; EU 1997) trigger of 10 for algae, whereas the TER value for *Lemna gibba*, based on the most sensitive EC50, falls below 10. Therefore, the potential risk of this herbicide to aquatic plants was evaluated further by consideration of the available higher-tier data.

### 4.2.3.2 Higher-Tier Data

In addition to the *Lemna gibba* study that was considered in the Tier 1 risk assessment, further laboratory data are available from a recovery study with this species. In this study, *Lemna* plants were exposed to the test substance for 4 or 7 days and subsequently transferred to untreated media for a further 7 days. Results from this study indicate that plants were able to recovery rapidly with EC50 values increasing to >3.8 and >9.4 µg ai/L, following 4- and 7-day exposure periods, respectively. In a further laboratory test, an additional 9 macrophyte species were exposed to the test substance for 7 days, followed by a 14-day recovery period. Endpoints were based on assessments of shoot length and weight. Endpoints from

Tier 1 and higher-tier studies were used to generate SSDs. All available endpoints were included, although EC50 values lying outside the exposure range of the test (i.e., greater-than values) were omitted. Results from this analysis indicated that *Lemna gibba* was the most sensitive species of those tested, and the median HC5 value was 1.43 µg ai/L.

The following data and its use were identified for comment:

1) applicability of initial or time-weighted average PEC values in the risk assessment,
2) consideration of recovery potential of *Lemna* and the additional species in the risk assessment, and
3) species selection (whether the selection of species adequately represents the range of macrophytes likely to be exposed in reality) and methods used in laboratory studies.

### 4.2.3.2.1    Applicability of Initial or Time-Weighted Average PEC Values in the Risk Assessment

The participants proposed that TER calculations based on time-weighted average PECs should use toxicity endpoints that have been calculated using time-weighted average concentrations based on measured concentrations in the test. Similarly, TER calculations based on maximum PECs should use toxicity endpoints based on maximum initial measured concentrations.

It was also recommended that there are 2 compound-specific characteristics that need to be considered in order to justify the use of time-weighted average concentrations: 1) the dissipation and degradation half-life and 2) the MoA and speed of action of the herbicide.

The participants considered that, for compounds with a rapid MoA, such as photosystem I (PSI) inhibitors, peak concentrations may be critical in determining the level of toxic response. Conversely, for compounds with a slower MoA, such as the sulfonylurea herbicides, the duration and concentration of exposure are more critical in determining the toxic response. In the FOCUS scenarios considered, PEC concentrations were approximately halved when using 7-day time-weighted averages compared to maximal concentrations in flowing waters, whereas there was almost no difference for static waters.

### 4.2.3.2.2    Consideration of Recovery Potential of Lemna and the Additional Species in the Risk Assessment

It was considered that the relevance of the recovery data to the risk assessment was dependent on the exposure profile and duration predicted in the FOCUS SW model. For static water bodies, where concentrations of the sulfonylurea were only expected to be halved after approximately 40 days, recovery data from a study with a 7-day exposure period was not considered applicable in a higher-tier assessment. In contrast, for flowing water bodies where herbicide concentrations are expected to decline more rapidly due to dilution, recovery data may be applicable. However, the workgroup noted that the route of herbicide entry into flowing water bodies also determined the relevance of recovery data because successive drainage or run-off events may prolong exposure periods, whereas a short pulse may be relatively short-lived. Consequently, the workgroup concluded that further information of the FOCUS SW exposure profiles is required in order to evaluate the relevance of recovery data in this particular case.

The workgroup also considered that the relevance of recovery data to the risk assessment was partly determined by the severity of the toxic effect caused by exposure to the PEC for realistic exposure durations. For example, toxic effects that lead to plant mortality would eliminate potential for recovery, whereas plants that suffer an inhibition of growth clearly have potential for recovery. In this case, all of the species tested showed some recovery after exposure to the herbicide for 7 days. If recovery of individuals or populations is to be tested, then observed recovery must be placed in an ecological context and must be able to be extrapolated to the field situation.

The workgroup concluded that in this particular case, recovery could not be considered in the higher-tier risk assessment due to the lack of information on the exposure profiles for each for the FOCUS SW scenarios and the concern that the 7-day exposure period did not reflect realistic exposure durations. Furthermore, participants felt that a 7-day exposure duration was not sufficient to detect effects in slower-growing species, particularly for sulfonylurea herbicides that are known to have a relatively slow MoA. However, the point was raised that the maximum PEC values occurred following applications to autumn-sown cereal crops in the D2 drainage scenario, which is largely found in the UK. A key issue is whether aquatic macrophyte species would be present in surface waters for exposure at that time of year, given that many species may die back or exhibit minimal growth during the winter months. No conclusion was reached on this point.

### 4.2.3.2.3   Species Selection and Test Methods Used in the Higher-Tier Laboratory Studies with Additional Species

From the evaluation of the higher-tier laboratory data, three issues related to their use in risk assessment were highlighted:

1) range of species in the SSD,
2) methodology used in the laboratory tests, and
3) number of species in the SSD.

The workgroup agreed that the test species represented a range of taxonomic groups and growth forms including rooted, submerged, emergent monocot, and dicot species. Hence, it was considered that the test adequately represented the range of macrophytes likely to be exposed in reality, but that the methodology used in the laboratory studies could be improved. In particular, a 7-day exposure period was not considered sufficient to allow for development of the full effects of compounds such as the sulfonylureas, particularly in light of the potential persistence of the herbicide in static waters.

The workgroup participants also discussed the options for the treatment of greater-than values in the SSD analysis and commented that the exclusion of these values from the analysis was akin to discarding one tail of the distribution and should be avoided. It was also recommended that in future studies, the concentration range of the test item should be extended to try to avoid the generation of greater-than values, unless the concentration range was limited by poor solubility or lack of effects in the test species at relatively high concentrations, that is, >100 mg ai/L.

# 5 Reports of Workgroups and Follow-Up Investigations

From the workshop, four areas worthy of further investigation were identified, and workgroups were established to continue development of knowledge and understanding. Workgroups are still discussing and developing their ideas, and hence these reports describe the objectives and preliminary results of each workgroup but should not be viewed as the final agreed outcome. The purpose of these reports is to generate interest and to provide the basis for further discussion and research. In addition it was recognized that there is a need to develop tools for spatio-temporal extrapolation of microcosm and mesocosm results. This activity would also be a topic for further research.

1) During the workshop, concern was expressed that *Lemna*, being a non-sediment-rooted monocot, may not be sensitive to residues in sediment or modes of action unique to dicot species. The need to evaluate the evidence for these concerns and develop decision-making criteria to determine when *Lemna* may not be an appropriate test species was recognized. Research to validate the need for additional testing was initiated. (Workgroup 1: Chair, Eric Bruns)

2) Workgroup participants acknowledged the requirement for an agreed test guideline for an alternative test species under circumstances where *Lemna* is not considered the most appropriate test species at Tier 1. For this purpose, a workgroup was established to develop and ring-test a protocol for an alternative test species, that is, *Myriophyllum* sp. (Workgroup 2: Chair, Peter Dohmen)

3) The lack of standardized test methods for macrophytes was acknowledged, and tasks were initiated to produce a database of existing methods based on the experience of participants and published literature. Information from experts via a questionnaire will be collated. (Workgroup 3: Chair, Peter Ebke)

4) Uncertainty was expressed over the use of macrophyte data in higher-tier assessments, specifically, the design and conduct of higher-tier studies with macrophytes and the use of macrophyte endpoints in SSDs. A task was initiated to develop guidance for the use of macrophyte data in SSDs (Workgroup 4: Chair, Stefania Loutseti)

## 5.1  WORKGROUP 1: CRITERIA FOR ASSESSING THE NEED FOR AN ADDITIONAL MACROPHYTE TEST

Chair: Eric Bruns
Members: Gertie Arts, Ute Kühnen

### 5.1.1  INTRODUCTION

At the AMRAP workshop, the current state of the European pesticide risk assessment concerning aquatic plants was discussed. In particular the potential need for additional ecotoxicological tests on aquatic macrophytes with herbicides was intensively discussed among scientists from regulatory authorities, business, and academia. The majority of the participants agreed on the need for a test with an additional aquatic macrophyte species to address the following concerns:

1) *Lemna*, as a floating, non-sediment-rooted monocotyledonous macrophyte, may not adequately represent all aquatic macrophytes. (It should be noted that algae tests also play a role in assessing the potential risk to aquatic plants.) In particular, *Lemna* sp. may not be a sensitive indicator for certain substances. Furthermore, there may be important differences between the sensitivity of monocotyledons, dicotyledons, or other taxonomic groups for different modes of action.

2) *Lemna* may not be sufficiently representative of rooted macrophytes and might not be suitable for assessing the risk from uptake of herbicides (from sediment) via roots.

While there was clearly concern that *Lemna,* together with the current algal testing scheme, may not be adequate surrogates for the entire aquatic plant community, there is little evidence to evaluate the extent to which these concerns have resulted in an inappropriate risk assessment. There are some published papers (Section 2.3) and unpublished information on the relative sensitivity of *Lemna* and other macrophytes to plant protection products. However, based on this rather limited information, it is not possible to state unequivocally that *Lemna* is less or more sensitive than other macrophyte species. The available information indicates that the existing testing scheme using *Lemna* and algae is generally sufficient to predict the lack of a significant risk to aquatic plants for a wide range of compounds. However, the proposed decision scheme shown in Figure 2.1 (Chapter 2), based on criteria below, has been developed to minimize underestimating the risk to the environment based on the use of algae and *Lemna* only. In this decision scheme, 3 criteria have been developed to address the issues identified above.

### 5.1.2  DECISION-MAKING CRITERIA FOR ADDITIONAL MACROPHYTE TESTS AT TIER 1

The criteria proposed in this paragraph are, where possible, based on either information or criteria presented in existing test guidelines and risk assessment guidance documents.

1) If information is available showing that for substances with a specific herbicidal MoA, algae or *Lemna* are less sensitive than other species (as has partly been shown for auxin-type herbicides), then a test with an additional macrophyte should be undertaken. Current knowledge concerning sensitivity differences of macrophytes to compounds with specific MoA, suggests that the risk for auxin-like compounds could possibly be underestimated by using *Lemna* only (Belgers et al. 2007). Therefore, compounds with an herbicidal MoA unique to dicot species should be subjected to an additional macrophyte test.

2) If an herbicidal or PGR compound is of low toxicity to algae and *Lemna*, and the EC50 values for these organisms are very high, then *Lemna* and algae may not be sufficiently sensitive to determine effects on primary producers. In these cases, a primary producer from another taxonomic group should be tested. Therefore, if the lowest EC50 values from algae and *Lemna* are greater than 1 mg/L, the compound should be tested in an additional macrophyte test.

3) Compounds that adsorb rapidly to the sediment layer may be less bioavailable for floating plants such as the Lemnaceae compared to rooted macrophytes. Therefore, testing (in addition) a rooted aquatic macrophyte should be considered if there is a significant likelihood that sediment will be the dominant exposure route; that is, if in water or sediment studies an appreciable amount (to be quantified) of the total applied substance is found in sediment at 7 or 14 days after application and the half-life in sediment is >14 d. In such a case, an additional macrophyte test should be performed unless it can be adequately demonstrated from terrestrial plant, crop residue, or other studies that the active substance does not exhibit activity via root uptake.

## 5.2   WORKGROUP 2: DEVELOPMENT OF A PROPOSED TEST METHOD FOR THE ROOTED AQUATIC MACROPHYTE, *MYRIOPHYLLUM* SP.

Chair: Peter Dohmen
Members: Gertie Arts, Eric Bruns, Nina Cedergreen, Jo Davies, Michael Dobbs, Peter Dohmen, Ute Feiler, Mark Hanson, Udo Hommen, Katja Knauer, Johanna Kubitza, Dirk Maletzki, Lorraine Maltby, Angela Poovey

### 5.2.1   INTRODUCTION

Standard test guidelines are available for algae (OECD 201) and *Lemna* (OECD 221), as a representative of higher aquatic plants, which can be used to generate data to address the risk of substances (and herbicides in particular) to aquatic non-target plant species. However, in some cases, these studies may not be sufficient, and information on an additional macrophyte may be required.

Based on current understanding and experience (Knauer et al. 2008; Kubitza and Dohmen 2008), the workgroup decided to focus on *Myriophyllum* (*M. spicatum* and/or *M. aquaticum*) as the additional macrophyte species. In the revision of 91/414/EEC (EU, 1997), *Myriophyllum sibiricum* is recommended in "a test which should preferably be conducted for auxin inhibitors and/or for compounds where data from terrestrial plants clearly demonstrate higher sensitivity of dicotyledonous plant species." Why *M. sibiricum* is mentioned, and not the far more intensively investigated and more frequently used *M. aquaticum* and/or *M. spicatum,* remains unclear to this group.

Here, a test method is proposed to assess the toxicity of substances to rooted aquatic plant species of the genus *Myriophyllum* (*M. aquaticum* and *M. spicatum*). The principal approach will also be useful for testing other aquatic macrophyte species, although some specific adaptations such as size of vessels, number of plants per replicate, and test duration may need modification. The method presented is based partly on existing guidelines OECD 221, 219, 201 (OECD 2006c, 2004b, 2006a), but includes modifications of those methods to reflect recent research and consultation on a number of key issues (Arts et al. 2008; Kubitza and Dohmen 2008; Knauer et al. 2008). The proposed method will be tested and validated by an international ring-test.

### 5.2.1.1 Principle of the Test

The objective of the test is to assess substance-related effects on the vegetative growth of the genus *Myriophyllum* in defined standard media (water, sediment, and nutrients) containing different concentrations of the test substance over certain test periods. For this purpose, individual shoot apices of healthy plants (without any flowers) potted in artificial standard sediment, containing additional nutrients, are maintained in a standard water medium. After an establishment period, the plants are exposed to a series of test concentrations added to the water column. The growth of the plants is evaluated for a period sufficient to allow a robust assessment of growth. At the end of the test, the plants are harvested and their biomass, length, and other relevant observations are recorded.

Biomass (whole-plant fresh weight) is the primary measurement variable. Additional measurement variables (such as dry weight, shoot length) are also measured. To quantify substance-related effects, growth in the test solutions is compared to that of the controls, and the concentration causing a specified percentage of inhibition of growth (e.g., 50%) is determined and expressed as the ECx (e.g., EC50).

### 5.2.1.2 Relevant Information on the Test Substance

The water solubility of the test substance, its vapor pressure, measured or calculated partitioning into sediment, and stability in water and sediment should be known. A reliable analytical method for the quantification of the test substance in water (and sediment) with known and reported accuracy and limit of detection should be available. Relevant information includes the structural formula and purity of the test substance. Chemical fate of the test substance (e.g., dissipation, abiotic and biotic degradation) may also be useful information.

### 5.2.1.3 Validity of the Test

Specific criteria will be set after the ring test. In the meantime, we propose that a minimum growth, such as a biomass increase in controls (>50%), growth supported throughout the test duration, and maintenance of temperature and pH within a pre-determined range, are considered as possible criteria. (Following pre-testing, it should be decided whether to include a quality criterion for unwanted algal contamination and which one.)

## 5.2.2 Description of the Test Method

### 5.2.2.1 Test Vessels

The study using *Myriophyllum* as test species is conducted in 2-L glass beakers (approximately 24 cm high and 11 cm in diameter). Other vessels may be suitable, but they should guarantee a suitable depth of water to keep the plants submersed throughout the study. Small plant pots (approximately 9 cm in diameter, 8 cm high, and 500 mL in volume) are used as containers for potting the plants into the sediment.

The sediment surface coverage should be >70% of the test vessel surface; the minimum overlaying water depth should be 12 cm.

### 5.2.2.2 Selection of Species

The method is designed to test selected *Myriophyllum* species, although previous studies indicate that it will probably be suitable for both *M. aquaticum* and *M. spicatum* (Kubitza and Dohmen 2008; Knauer et al. 2008). Species identification must be verified (in North America there is evidence of hybridization between *M. spicatum* and related species; Moody and Les 2002), and the source of the plants should be described.

The plants should first be kept in the laboratory and be visibly free of any other species (particularly snails or filamentous algae; in some regions eggs or larvae from the small moth *Paraponyx stratiotata* also can be a problem; some level of epiphytes — such as diatoms, but no filamentous algae — may often be unavoidable and will generally not be a problem). Only visibly healthy plants, without flowering shoots, should be used for the study.

If the plants are kept within the laboratory before the test as a maintenance culture, then temperature, light, and nutrient conditions should be at the low end, that is, nutrient concentrations reflecting oligotrophic to mesotrophic systems. For this, tap water may be useful. If culturing or maintenance conditions differ significantly from lab conditions (i.e., if plants are taken from outdoor systems at a time when temperature and day length differ significantly from those in the lab), then plants should be cultured before the study under conditions similar to the study in order to support good growth and allow acclimatization.

### 5.2.2.3 Sediment

The following formulated sediment, based on the artificial soil used in OECD Guideline 219, is recommended for use in this test:

1) 4% to 5% peat (dry weight, according to $2 \pm 0.5\%$ organic carbon) as close to pH 5.5 to 6.0 as possible; it is important to use peat in powder form, finely ground (particle size < 1 mm) and only air dried.
2) 20% (dry weight) kaolin clay (kaolinite content preferably above 30%).

3) 75% to 76% (dry weight) quartz sand (fine sand should predominate with
more than 50% of the particles between 50 and 200 µm).

4) Instead of deionised water, aqueous nutrient medium (N/P source) is added
to obtain moisture of the final mixture of about 40% (details will be clari-
fied after pre-testing).

5) If needed, calcium carbonate of chemically pure quality ($CaCO_3$) is added
to adjust the pH of the final mixture of the sediment to $7.0 \pm 0.5$.

The source of peat, kaolin clay, and sand should be known and documented. If
the origin is unknown or gives concern, then the respective components should be
checked for the absence of chemical contamination (e.g., heavy metals). It may be
useful to add a fine, very thin layer of quartz sand on top of the sediment to reduce
suspension of sediment into the water.

### 5.2.2.4  Water Medium

Different media have been tested for culturing and testing *Myriophyllum* such as
Smart and Barko (1985), Elendt M4 (Elendt 1990; macro elements only), or AAP
medium (OECD 2002). (Final recommendations will be given after pre- and ring-
testing.)

The pH at test initiation should be between 7.5 and 8.0 to allow for optimum
plant growth.

### 5.2.2.5  Test Procedure

Healthy shoot apices from the culture plants are cut off at a length of 6 cm ($\pm$ 0.5 cm).
These shoot tips are maintained prior to the test for either 3 days for *M. aquaticum*
or 7 days for *M. spicatum*, in culture vessels with the lower 3 cm, including 2 nodes,
in the sediment overlaid with nutrient-poor water (or just the lower part in a basal
medium for *M. aquaticum*) to induce root development.

Thereafter, for *M. aquaticum* plants are removed from the pre-culture and cleaned
of sediment and surplus water; plants that are apparently not healthy or have not
developed any roots will be discarded at this stage. The plants are weighed (to reduce
variability, the weight of the shoot tips used in the study should not differ by more
than 30% from the mean). Shoots are then potted into the sediment as before, and
shoot length above sediment is measured.

For *M. spicatum,* the pre-adaption starts with 5 plants per pot. After this period, 2
plants are removed to remain with 3 largely homogeneously performing individuals.
Five additional pots are harvested at this stage, and shoot length (eventually including side shoots) and plant biomass (weight) are monitored. The pots are transferred
to vessels that will be filled with fresh test solutions.

Five plants (3 for *M. spicatum*) are used per test vessel, and 3 replicates are prepared for each treatment group (in general, 5 test concentrations arranged in a geo-
metric series) plus 5 replicates for the control. (More plants per pot are used for *M.
aquaticum*, which has a slightly different growth form to *M. spicatum* and allows
more plants per pot.) The individual plant shoots should be added impartially to the
test vessels, and the test vessels should be impartially (randomly) assigned to the
different treatment groups.

The pots with sediment and plants are placed into the glass beakers. Afterwards, the test vessels are carefully filled with the respective amount of medium containing the relevant amounts of the test substance. If the test substance is added afterwards, it should be done in a way that guarantees a homogeneous distribution within the test system. The final water volume (1.8 L has been proven to work for the kind of test vessels described above) must be known and the concentrations are set up accordingly. If water evaporates during the test by more than 10% (it may be helpful to mark the water level at test initiation), the water level should be raised with distilled water.

The exposure period should be 7 days for *M. aquaticum* and 14 days for *M. spicatum*. (For compounds known to show a slow or delayed response, it may be appropriate to increase the test duration by 1 week; the growth rate over time may indicate such a delayed response.) During this time, shoot length and any other observations are recorded at least twice during the exposure period (i.e., on days 3 and 5 for *M. aquaticum* and on days 5 and 10 for *M. spicatum*). Shoot length (main and side shoots) is determined, for example, using a ruler positioned within the vessel close to the plant to be measured. It may be necessary to straighten shoots a bit (obviously this needs to be done without inflicting any damage to the plant) for more accurate length measurements.

At the end of the test, all plants are measured again, and any growth anomalies are recorded; thereafter, the whole plants are harvested. Any symptoms (such as chlorosis or necrosis) or other observations are recorded. Total plant wet weight (after the remaining test medium is carefully blotted off) and subsequently total plant dry weight are determined; the necessity of this latter parameter will be decided after ring-testing. (Most information available indicates that — particularly for plants — dry weight measurements are of very limited value. The need for this parameter should thus be discussed, eventually after the ring-test results.) A visual assessment of the roots is made and any unusual findings should be recorded. If side shoots are present, their numbers and length should also be measured.

Light conditions, pH, oxygen levels, and water temperature are determined at test initiation. Temperature in water (and/or within the room) should be monitored over the whole test period. The pH and oxygen concentration of the test medium (water) should be checked at test initiation, at least once during the study (every 3 to 4 days) and at the end of the study in all replicate vessels.

A randomized design for the location of the test vessels in the growth chamber is required to minimize the influence of spatial differences in light intensity or temperature. A respective blocked design or random repositioning of the vessels needs also to be taken into account after observations are made.

### 5.2.2.6  Test Conditions

Warm and/or cool white fluorescent lighting should be used to provide a light intensity selected from the range of about 100 to 120 $\mu E \cdot m^{-2} \cdot s^{-1}$ when measured in a photosynthetically active radiation (400 to 700 nm) (equivalent to about 8 000 lux) at the water surface and using a light-to-dark ratio of 16:8 h. (The method of light detection and measurement, in particular the type of sensor, will affect the measured value. Spherical sensors, which respond to light from all angles above and below the plane of measurement, and "cosine" sensors, which respond to light from all angles above the plane of measurement, are preferred to unidirectional sensors and will give higher readings for a multi-point light source of the type described above.) Any differences from the selected light intensity over the test area should not exceed the range of ±15%. The temperature in the test vessels should be 20 ± 2 °C. The pH of the control medium should not increase by more than 1.5 units during the test (this value may need to be modified depending on ring-test results). However, deviation of more than 1.5 units would not invalidate the test when it can be shown that validity criteria are met.

### 5.2.2.7   Analytical Measurements of Test Substance

The correct application of the test substance must be supported by analytical measurements of substance concentrations in water at test initiation.

In addition, test substance concentrations in water must be checked at test termination (where appropriate in long term tests also at an intermediate time interval), which may help in the interpretation of the results together with information from other test systems (such as water and sediment studies).

### 5.2.2.8   Data Evaluation

The purpose of the test is to determine the effects of the test substance on the vegetative growth of the test species. The average specific growth rate for a specific period is calculated as the logarithmic increase in the growth variables — plant wet (and dry) weight, shoot length, numbers of side shoots, and one other measurement variable — using the formula below for each replicate of control and treatments:

$$\mu_{i\text{-}j} = \ln(N_j) - \ln(N_i) \,/\, t$$

where $\mu_{i\text{-}j}$ is the average specific growth rate from time $i$ to time $j$, $N_i$ is the measurement variable in the test or control vessel at time $i$, $N_j$ is the measurement variable in the test or control vessel at time $j$, and $t$ is the time period from $i$ to $j$.

For each treatment group and control group, a mean value for growth rate along with variance estimates should be calculated.

The average specific growth rate should be calculated for the entire test period (time $i$ in the above formula is the beginning of the test and time $j$ is the end of the test). For each test concentration and control, a mean value for average specific growth rate along with the variance estimates should be calculated. In addition, the interim growth rates during the exposure period should be assessed based on shoot length data in order to evaluate effects of the test substance occurring during the exposure period and to check whether sufficient growth occurred throughout the exposure period.

Percent inhibition of growth rate ($I_r$) may then be calculated for each test concentration (treatment group) according to the following formula:

$$\%I_r = \frac{\mu_C - \mu_T}{\mu_C} \times 100$$

where: $\%I_r$ is the percent inhibition in average specific growth rate, $\mu_C$ is the mean value for $\mu$ in the control, and $\mu_T$ is the mean value for $\mu$ in the treatment group.

### 5.2.2.8.1 Plotting Concentration–Response Curves

Concentration–response curves relating mean percentage inhibition of the response variable ($I_r$), calculated as shown above, and the log concentration of the test substance should be plotted.

### 5.2.2.8.2 ECx Estimation

Estimates of the ECx (e.g., EC50 and EC20) should be based upon average specific growth rates ($E_rC_x$), which should in turn be based upon biomass data and, where relevant, upon other additional measurement variables.

### 5.2.2.8.3 Statistical Procedures

The aim of the study is to obtain a quantitative concentration–response relationship, which is usually done by regression analysis. It is possible to use a weighted linear regression after having performed a linearizing transformation of the response data, for instance into probit or logit or Weibull units, but nonlinear regression procedures are preferred techniques that better handle unavoidable data irregularities and deviations from smooth distributions. Approaching either zero or total inhibition, such irregularities may be magnified by the transformation, interfering with the analysis. Standard methods of analysis using probit, logit, or Weibull transforms are intended for use on quantal (e.g., mortality or survival) data, and should be modified to accommodate growth rate or yield data.

For each response variable to be analyzed, the concentration–response relationship should be used to calculate point estimates of ECx values. When possible, the 95% confidence limits for each estimate should be determined. Goodness of fit of the response data to the regression model should be assessed either graphically or statistically. Regression analysis preferably should be performed using individual replicate responses, not treatment group means.

EC50 estimates and confidence limits may also be obtained using linear interpolation with bootstrapping, if available regression models or methods are unsuitable for the data.

## 5.2.3 REPORTING

The test report must include the following:

1) Test substance
2) Test species
   a. scientific name and source
   b. description of culture or field population from which stock has been derived

3) Test conditions
   a.  test procedure used
   b.  date of start of the test and its duration
   c.  test medium
   d.  description of the experimental design: test vessels and covers, solu-
       tion volumes, number of shoots per test vessel at the beginning of
       the test
   e.  test concentrations (nominal and measured as appropriate) and number
       of replicates per concentration
   f.  methods of preparation of stock and test solutions, including the use of
       any solvents or dispersants
   g.  temperature during the test
   h.  light source, light intensity and homogeneity
   i.  pH values of the test and control media
   j.  methods for determination response variables, for example, dry
       weight, fresh weight
   k.  all deviations from this procedure

4) Results
   a.  raw data: number, weight, length of shoots, and other measurement vari-
       ables in each test, and control vessel at each observation and occasion of
       analysis
   b.  means and standard deviations for each measurement variable; growth
       curves for each concentration (recommended with log-transformed
       measurement variable)
   c.  growth rate in the controls throughout the exposure period based on
       shoot length
   d.  calculated response variables for each treatment replicate, with mean
       values and coefficient of variation for replicates
   e.  graphical representation of the concentration–effect relationship
   f.  estimates of toxic endpoints for response variables for example, EC50,
       EC20, and associated confidence intervals
   g.  any stimulation of growth found in any treatment
   h.  any visual signs of phytotoxicity, as well as observations of test
       solutions
   i.  discussion of the results

## 5.2.4  APPENDIX TO CHAPTER 5: NUTRIENT MEDIA

The following (Table 5.1) describes the composition of AAP medium as used in
*Lemna* testing.

Add the nutrient stock solutions in the amounts indicated and in the order indicated to approximately 900 mL of distilled or deionized water; then adjust the final
volume after pH adjustment to 1 L. The pH of the medium should be adjusted, as
necessary, to 7.5 ± 0.1 using 0.1 N NaOH or 10% HCl.

The first publication of the M4 medium can be found in Elendt (1990). Data indicate that a restriction to the macro elements, together with some fertilizer added to the sediment, works sufficiently well without the need for the M4 micronutrients or vitamins.

## TABLE 5.1
### Preparation of 1x AAP medium

| Constituent | Stock Solution Preparation | Amount of Each Stock Solution to Add to Prepare 1 L of 1x AAP | Final Concentration in Medium |
|---|---|---|---|
| NaNO$_3$ | 12.750 g/ 0.5 L | 1 mL/L | 25.50 mg/L |
| MgCl$_2$• 6 H$_2$O | 6.082 g/ 0.5 L | 1 mL/L | 12.16 mg/L |
| CaCl$_2$ • 2 H$_2$O | 2.205 g/ 0.5 L | 1 mL/L | 4.41 mg/L |
| Micronutrient Stock Solution | Add all of the following to 0.5 L of water | 1 mL/L | |
| H$_3$BO$_3$ | 92.760 mg | | 0.1855 mg/L |
| MnCl$_2$• 4 H$_2$O | 207.690 mg | | 0.415 mg/L |
| ZnCl$_2$ | 1.635 mg | | 3.27 µg/L |
| FeCl$_3$• 6 H$_2$O | 79.880 mg | | 0.1598 mg/L |
| CoCl$_2$• 6 H$_2$O | 0.714 mg | | 1.428 µg/L |
| NaMoO$_4$• 2 H$_2$O | 3.630 mg | | 7.26 µg/L |
| CuCl$_2$• 2 H$_2$O | 0.006 mg | | 0.012 µg/L |
| Na$_2$EDTA• 2 H$_2$O | 150.000 mg | | 0.300 mg/L |
| MgSO$_4$• 7 H$_2$O | 7.350 g/ 0.5 L | 1 mL/L | 14.70 mg/L |
| K$_2$H PO$_4$ | 0.522 g / 0.5 L | 1 mL/L | 1.044 mg/L |
| NaHCO$_3$ | 7.500 g/ 0.5 L | 1 mL/L | 15.00 mg/L |

## TABLE 5.2
### Preparation of Smart and Barko medium (1985)

| Constituent | Amount of Reagent Added to Water |
|---|---|
| CaCl$_2$ • 2 H$_2$O | 91.7 mg/L |
| MgSO$_4$• 7 H$_2$O | 69.0 mg/L |
| NaHCO$_3$ | 58.4 mg/L |
| KHCO$_3$ | 15.4 mg/L |

*pH (air equilibrium) – 7.9*

**TABLE 5.3**
**Preparation of M4 medium**

| Macro Nutrient Stock Solutions (Single Substance) | Amount Added to Water [mg/L] | Concentration (Related to Final Medium M4) | Amount of Stock Solution Added to Prepare Medium [mL/L] |
|---|---|---|---|
| $CaCl_2 \cdot 2\,H_2O$ | 293 800 | 1 000-fold | 1.0 |
| $MgSO_4 \cdot 7\,H_2O$ | 246 600 | 2 000-fold | 0.5 |
| KCl | 58 000 | 10 000-fold | 0.1 |
| $NaHCO_3$ | 364 800 | 1 000-fold | 1.0 |
| $Na_2SiO_3 \cdot 9\,H_2O$ | 50 000 | 5 000-fold | 0.2 |
| $NaNO_3$ | 2 740 | 10 000-fold | 0.1 |
| $KH_2PO_4$ | 1 430 | 10 000-fold | 0.1 |
| $K_2HPO_4$ | 1 840 | 10 000-fold | 0.1 |

## 5.3 WORKGROUP 3: USE OF ADDITIONAL MACROPHYTE TEST SPECIES: CURRENT EXPERIENCE

Chair: Klaus Peter Ebke
Members: Gertie Arts, Katie Barrett, Nina Cedergreen, Heino
Christl, Jo Davies, László Dören, Udo Hommen, Katja Knauer,
Johanna Kubitza, Jonathan Newman, Petra Pucelik-Günther

One outcome of the AMRAP workshop was the recognition that standardized methods for testing pesticide effects on macrophyte species other than *Lemna* sp. are not available in Europe. Therefore, a working group was established with the aim of producing a list of additional test species considering taxonomy, growth form, availability, test duration (based on growth), and experience on culturing and handling.

The working group started to collect information on experience with laboratory single-species tests with macrophytes other than *Lemna* sp. The questionnaire "Experience with macrophytes in ecotoxicology" was developed and distributed to a circle of experts in Europe. The questions in the form were related to the species of macrophytes that have been tested, the test conditions (water, sediment, test duration), the endpoints measured and calculated, the validity criteria applied, the source of the test macrophytes, and other details considered of importance for the individual studies.

To date, 10 institutions have responded to the questionnaire (Table 5.4). In total, information on 96 tests was provided, covering 28 different species of aquatic macrophytes (Table 5.5).

Thirty-seven tests were conducted as flow-through tests, whereas 59 tests were static tests. Forty-five tests were conducted with only an aqueous medium, whereas 42 tests included sediment. The most commonly measured parameters were length, (fresh-) weight, and leaf area. Length, weight or area measurements, and other endpoints used for the evaluation of toxicity might give different interpretations of sensitivity; for example, in a single-species test, the value of EC50 of biomass may not be the same as the value of EC50 of length-specific growth rate. In some tests, root endpoints were included, and in only a very few tests, physiological parameters such as photosynthetic efficiency were included. The test durations varied between 1 and 4 weeks.

In most cases, validity criteria were not defined. Some researchers suggest a minimum growth rate of the control plants, depending on the test species. The suggestions varied from 30% to 100%, depending on growth endpoints (e.g., length or leaf area).

The aim of the ongoing evaluation of the questionnaire "Experience with macrophytes in ecotoxicology" is to give more detailed and quantitative answers to the following questions:

- Which species can be grown successfully under laboratory conditions?
- What are optimum growth conditions for different species in terms of light, nutrient, media, temperature, etc.?
- What are the maximum growth rates we can expect under laboratory conditions for different species?
- How relevant are further endpoints (e.g., root growth or physiological parameters)?
- Which substances can be used as reference substances in testing macrophytes?

The ongoing evaluation should increase the database. Scientists working on macrophytes are invited to submit their experience by the questionnaire (contact the authors for a copy of the file). Once additional data are available, these will be used to update and expand Tables 5.3.1 and 5.3.2 below.

**TABLE 5.4**
**List of the institutions that responded to the questionnaire**

Alterra, WUR
University of Basel, Dept. of Environmental Sciences
BASF SE
BfG Koblenz
University of Copenhagen, Dept. of Agricultural Sciences
Federal Office of Agriculture
Fraunhofer IME
MESOCOSM GmbH
RCC Ltd
UBA (Federal Environment Agency Germany)

**TABLE 5.5**
**List of aquatic macrophytes used by institutions that responded to the questionnaire**

| Species | Number of Data Sets | Distribution (in Relation to Europe) | Taxonomy | Growth or Life-Form | Rooting in Sediment |
|---|---|---|---|---|---|
| *Berula erecta* | 1 | native | dicotyledonous | emergent | yes |
| *Callitriche palustris* | 1 | native | dicotyledonous | submerged | yes |
| *Callitriche plathycarpa* | 3 | native | dicotyledonous | submerged | yes |
| *Ceratophyllum demersum* | 11 | native | dicotyledonous | submerged | no |
| *Ceratophyllum submersum* | 3 | native | dicotyledonous | submerged | no |
| *Chara intermedia* | 2 | native | macro-algae | submerged | no |
| *Crassula helmsii* | 1 | invasive plant originally from Australia, New Zealand | dicotyledonous | submerged | yes |
| *Egeria densa* | 5 | invasive plant originally from South America | monocotyledonous | submerged | yes |
| *Elodea canadensis* | 10 | invasive plant originally from North America | monocotyledonous | submerged | yes |
| *Elodea nuttallii* | 4 | invasive plant originally from North America | monocotyledonous | submerged | yes |
| *Glyceria maxima* | 2 | native | monocotyledonous | emergent | yes |
| *Heteranthera zosterifolia* | 2 | aquarium plant from South America | monocotyledonous | submerged | yes |
| *Hygrophila polysperma* | 2 | aquarium plant from India and Bhutan | dicotyledonous | submerged – semi-emerged | yes |
| *Lemna trisulca* | 3 | native | monocotyledonous | free-floating | no |
| *Myriophyllum aquaticum* | 8 | invasive plant originally from South America | dicotyledonous | submerged – semi-emerged | yes |
| *Myriophyllum heterophyllum* | 1 | invasive plant originally from North America | dicotyledonous | submerged | yes |

(*Continued*)

**TABLE 5.5 (Continued)**

| Species | Number of Data Sets | Distribution (in Relation to Europe) | Taxonomy | Growth or Life-Form | Rooting in Sediment |
|---|---|---|---|---|---|
| *Myriophyllum spicatum* | 11 | native | dicotyledonous | submerged | yes |
| *Myriophyllum verticillatum* | 1 | native | dicotyledonous | submerged | yes |
| *Potamogeton crispus* | 7 | native | monocotyledonous | submerged | yes |
| *Potamogeton natans* | 2 | native | monocotyledonous | submerged and floating leaf | yes |
| *Ranunculus aquatilis* | 1 | native | dicotyledonous | submerged and floating leaf | yes |
| *Ranunculus circinatus* | 1 | native | dicotyledonous | submerged | yes |
| *Ranunculus trichophyllus* | 1 | native | dicotyledonous | submerged | yes |
| *Riccia fluitans* | 2 | native | Bryophyta | submerged | no |
| *Sparganium emersum* | 4 | native | monocotyledonous | emergent | yes |
| *Spirodela polyrhiza* | 4 | native | monocotyledonous | free-floating | no |
| *Stratiotes aloides* | 1 | native | monocotyledonous | submerged | no |
| *Vallisneria spiralis* | 2 | invasive plant originally from subtropic and tropic regions | monocotyledonous | submerged | yes |

## 5.4 WORKGROUP 4: INVESTIGATING THE SSD APPROACH AS A HIGHER-TIER TOOL FOR RISK ASSESSMENT OF AQUATIC MACROPHYTES

Chair: Stefania Loutseti
Members: Gertie Arts, Nina Cedergreen, Heino Christl, Jo Davies, Peter Dohmen, Mick Hamer, Mark Hanson, Joy Honegger, Lorraine Maltby, Melissa Reed

### 5.4.1 BACKGROUND AND OBJECTIVES

The SSD workgroup was formed at the AMRAP workshop to address specific scientific and regulatory questions on the use of SSDs as a higher-tier risk assessment tool for aquatic macrophytes and was charged with addressing the following:

1. Species selection for SSDs
   • What is the sensitivity of *Lemna* species relative to other macrophytes?

2. Endpoint selection for SSDs
   • Which endpoints should be included in an SSD?
   • Should SSDs be based on a common endpoint or the most sensitive endpoint for each species?
   • What is the range in endpoint sensitivity within and between species?

During the course of preliminary discussions, the following 2 issues were also raised by the workgroup as possible objectives:

3. Can algae and macrophyte data be combined in the same SSD?
4. How should "greater-than" values be incorporated in the SSD approach?

There was agreement on addressing questions 1 and 2, but not on questions 3 and 4, which are therefore still outstanding.


## 5.4.2 Creation of a Database on Aquatic Macrophyte Toxicity

In order to answer these questions, a database of aquatic plant endpoints was created from several sources of information. Between February and May 2008, the following contributors collated macrophyte data into the database:

1) Dr. Gertie Arts, Alterra
   The data sources for the database are scientific reports, documents, and papers that provide sufficient information on the experimental setup of the toxicity tests (exposure time, laboratory or (semi-)field study, water regime used, etc.). References were selected if published between 1980 and 2008. All references were checked. Publications were extracted from Wageningen UR library, in both hard copy and digital format. All Alterra macrophyte toxicity data (published and unpublished nonconfidential data, used with permission) are included in the database. Part of the older data was already described in De Zwart (2002).

2) Dr. N. Cedergreen, DK University
   Data were delivered from the following studies: Cedergreen et al. 2004a, 2004b, and 2005. For the study on metsulfuron-methyl with *Elodea canadensis*, measured endpoints other than the published ones were also included in the database.

3) Dr. Udo Hommen, Fraunhofer IME
   Data from confidential GLP studies of the Fraunhofer IME (used with permission) to construct macrophyte SSDs for 2 herbicides were provided. In addition, metazachlor toxicity data for macrophyte species were taken from a poster of Kubitza and Dohmen (BASF) presented during the 2002 SETAC Europe conference in Vienna.

4) Dr. Heino Christl, JSC International Ltd

Data were generated in several non-GLP pilot studies (but in a GLP environment). Two of them were performed with a test substance of known properties. The aim was to establish and develop methods that would allow regulatory issues with aquatic macrophytes to be addressed.

The information in Table 5.6 was collated for each endpoint.
The toxic mode of action (TMA) was differentiated as shown in Table 5.7.
Each endpoint was designated according to 3 descriptors:

- the statistical endpoint, that is, EC50 or NOEC;
- the plant part, that is, total plant, shoot, root, or frond; and
- the measurement type, that is, e.g., dry weight (increase), number (final), length (final), length (increase).

Each set of endpoints that was derived from one experiment or set of test plants was identified by a unique study number.

## TABLE 5.6
## Data on aquatic macrophyte toxicity endpoints

| Field | Explanation or Examples |
|---|---|
| Source of database | Organization or researcher |
| Type of pesticide | Herbicide, fungicide, degradation product |
| CAS# | CAS number |
| Chemical name or family of molecular formula | Chemical name |
| TMA | Toxic mode of action, see Table 5.7 |
| Syn1 | Name of compound |
| Test species | Latin name |
| Group | Macrophytes |
| Endpoint | EC50, EC10, NOEC |
| Plant part | Root, shoot, leaf, frond, total plant, etc. |
| Assessment | Measurement endpoint, e.g., final weight, dry weight, wet weight, length increase |
| Test number | Code to indentify data from one test |
| Type of study | Lab |
| Class | Taxonomic class |
| Family | Taxonomic family |
| Plant morphology | Floating, submerged, emergent |

*(Continued)*

**TABLE 5.6    (Continued)**

| Field | Explanation or Examples |
|---|---|
| Rooted in sediment | Rooted or not |
| Exposure regime | Static, semi-static, flow-through |
| Duration days | Test duration in days |
| Conc μg/L | Concentration in μg a.s./L |
| Reference | Published paper or report |
| Remarks | Remarks on test solution, sediment type, data used for SSDs |

**TABLE 5.7**
**Toxic modes of action**

| Code | Toxic Mode of Action |
|---|---|
| IFR | Inhibits fungi respiration |
| IFG | Inhibits fungi growth |
| IAMS | Inhibits amino acids synthesis |
| AUXS | Simulates auxin hormone |
| IPS | Inhibits photosynthesis |
| EDPSI | Electron diversion PS I |
| ICD | Inhibits cell division |
| FAZ | Formation of abscission zone |
| IFAS | Inhibits fatty acids synthesis |
| IRFE | Inhibits root formation or elongation |
| ICDE | Inhibits cell division or elongation |
| OPG | Obstruct plant growth |
| IMB | Inhibits multiple biosyntheses |
| ICAC | Inhibits citric acid cycle |
| EPSPI | EPSP synthase inhibitor |

### 5.4.3  PROGRESS TO DATE AND WAY FORWARD

To date, data representing more than 2000 endpoints for 54 compounds, predominantly herbicides, in 55 freshwater aquatic macrophyte species, have been added to the database. The working group has also undertaken a preliminary statistical analysis of the database. During the course of this analysis, it became apparent that the composition of the database and level of description for each endpoint was insufficient to allow a robust statistical analysis and interpretation. Consequently, the workgroup is continuing to develop the database, and the following actions are now necessary in order to complete this exercise and to define quality criteria for the inclusion of endpoints in the database:

- Audit the database based on these quality criteria.
- Extend the database to include standard *Lemna* endpoints in order to answer Question 1 (sensitivity of *Lemna* compared to other macrophytes).
- Extend the database by inviting a broader audience (i.e., industry colleagues) to contribute unpublished company data.
- Consider the inclusion of algal endpoints into the database in order to address Objective 3.
- Develop a defined protocol for completing the statistical analysis of the database and the analysis of the data.

# 6 Keynote Presentations

Keynote presentations are provided as information sources and were used at the AMRAP workshop to provide an introduction and thought-starter for the subsequent discussions. There are 5 presentations, each focusing on different issues concerning aquatic macrophytes. Elements from them were developed during workshop discussions.

## 6.1  AQUATIC MACROPHYTES IN AGRICULTURAL LANDSCAPES

Jeremy Biggs, Pond Conservation, UK

### 6.1.1  INTRODUCTION

Aquatic macrophytes are important in freshwaters in the agricultural landscape for 3 main reasons. First, they are part of biodiversity and have an inherent value in their own right. Second, they support other biodiversity, both animals and plants, and third, they carry out a range of ecosystem functions that help to maintain the integrity of freshwater systems. In this paper, these functions are briefly reviewed and the composition of macrophyte assemblages in different waterbody types (rivers, streams, ponds, ditches) in the agricultural landscape is described.

### 6.1.2  FUNCTIONAL ROLE OF MACROPHYTES

#### 6.1.2.1  Wetland Plants as a Component of Biodiversity

Although macrophytes are a large and conspicuous part of freshwater ecosystems, in simple terms of numbers of species, they represent a relatively small proportion of the total aquatic biodiversity. For example, in the UK, there are about 350 wetland plant species, which represent in the order of 5% of all freshwater species (excluding bacteria and fungi) (Figure 6.1).

In this chapter, the distribution patterns of aquatic macrophytes are analyzed using groupings based on their morphology: emergent species, floating-leaved species (rooted and free-floating), and submerged aquatic species. Although these broad categories are useful in describing the growth form of the plants, they do not provide a precise indication of their habitats. For example, many submerged aquatic species can live in very shallow water, including the drawdown zone, especially in smaller water bodies typical of agricultural landscapes.

There is clear evidence that many native European wetland plant species have become less widespread and abundant, particularly because of eutrophication. For

65

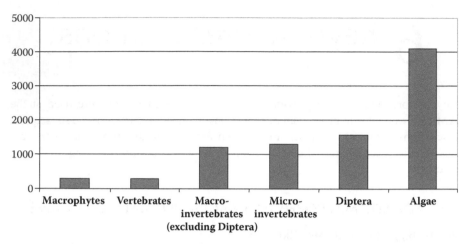

**FIGURE 6.1**  Approximate species richness of major freshwater groups in United Kingdom. (Source: Pond Conservation databases with permission.)

example, in the UK the majority of lakes have changed very considerably from their baseline chemical quality, with one study showing nearly half (46%) having a 500% increase in nutrient loading compared to the predicted natural baseline (Jones et al. 1998). Associated with this change (although not exclusively explained by it), there have been large contractions in the range of many nutrient-sensitive wetland plants. In contrast to the large declines seen in native species, a small number of species not native to Europe have become more widespread. In the UK, these species include the introduced North American *Elodea* species and the Australasian *Crassula helmsii* (Preston et al. 2002). Data from ponds, which are often intimately associated with the farmed environment because of their small catchments (Davies et al. 2008), give an indication of the specific level of impact on wetland macrophytes associated with agriculture. Thus ponds in the British landscape on average support about half the number of wetland plant species expected, compared to minimally impaired reference conditions. Losses are particularly severe among the submerged plants, where typically impacted farmland ponds support only 30% of the expected species (Biggs et al. 2005).

### 6.1.2.2   Macrophytes as Habitat for Other Organisms
Macrophytes provide habitats for other plants, particularly epiphytic algae and animals. Typically, macrophytes add to the variety of habitats available for animals (White and Irvine 2003) and, in some types of macrophyte stands, increase the abundance of invertebrates compared to more open habitats. For example, in gravel pit lakes in southern England, unvegetated gravel and sand supported an average of about 10 macroinvertebrate species in each hand net sample of 10 s duration, compared with vegetated habitat, which supported nearly double this number of species (Figure 6.2).

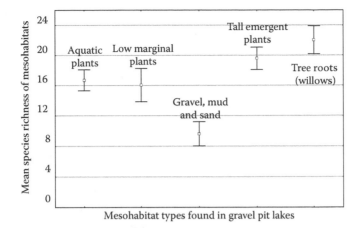

Mesohabitat types found in gravel pit lakes

**FIGURE 6.2** Comparative macroinvertebrate species richness of different mesohabitats with and without macrophytes in gravel pit lakes in southern England (Pond Conservation unpublished data with permission). The graph is based on analysis of macroinvertebrate data from 11 lakes with replicate hand-net samples from 5 mesohabitat types ($n = 502$). The study was carried out in 1990 and 1991 as part of a project assessing the nature conservation value of gravel pit lakes.

It is also clear that different macrophyte types support different macroinvertebrate assemblage types, indicating that macrophytes have an important role in generating biological diversity. For example, in gravel pit lakes in southern England, multivariate analysis of the macroinvertebrate assemblages of different mesohabitat types showed that distinct differences were apparent between different mesohabitat types (usually different kinds of vegetation stand). Invertebrate samples from the same mesohabitat were generally more similar to each other than to those of different mesohabitat types (Figure 6.3).

Patterns in macroinvertebrates on different vegetation types have been reasonably well studied but appear to be similar among microinvertebrates. For example, recent data on ciliates show greater number and biomass of animals among submerged plants, followed by emergent macrophytes, with the lowest numbers in open water (Mieczan 2007).

### 6.1.2.3 Macrophytes as Food and Oviposition Sites

Aquatic plants provide food for invertebrates and vertebrates (birds and fish), although once dead and decaying, they are mostly consumed by invertebrates. Macrophytes also provide a very important substrate on which epiphytic algae can grow, which are an important live food source for animals. Although it has been known for a long time that birds eat aquatic plants, it is increasingly apparent that birds, especially waterfowl, can have a big impact on macrophytes. Normally, it is assumed that this effect is most pronounced in lakes and ponds, but there is also recent evidence of birds affecting vegetation abundance in running waters. For example, in a shallow

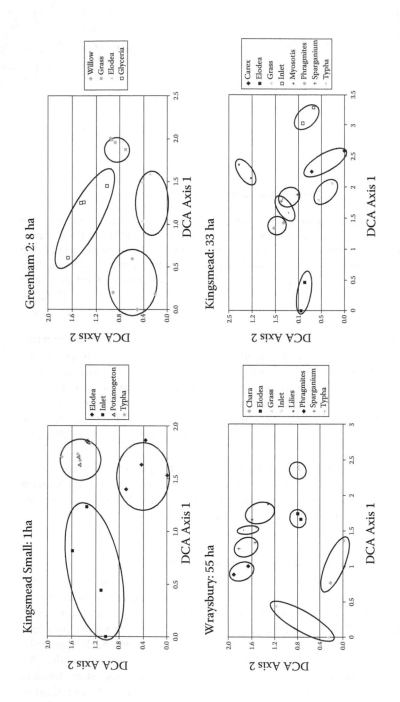

**FIGURE 6.3** Variation in macroinvertebrate species assemblages in different mesohabitats in 4 gravel pit lakes in southern England. (Ellipses link samples from the same mesohabitat.) Invertebrate samples from the same mesohabitat were generally more similar to each other than to those of different mesohabitat types. The graph is based on Pond Conservation's unpublished data with permission from a study of macroinvertebrate habitats in 11 lakes from which $n = 160$ replicated mesohabitat samples were collected.

## TABLE 6.1

## Examples of the use of macrophytes as oviposition sites by invertebrates

| Plant Sites | Invertebrates |
| --- | --- |
| Submerged and floating-leaved plants (including *Lemna*) | Water snails |
| | Damselflies, some dragonflies |
| | Water bugs (e.g., *Notonecta, Nepa*) |
| | Aquatic moths |
| | Some caddis flies |
| | Some jewel beetles (e.g., *Donacia crassipes* with lilies) |
| | Weevils: *Tanysphyrus lemnae*, the Duckweed Weevil |
| Emergent plants: | Jewel beetles (*Donacia* spp.) |
| | Moths (e.g., wainscot moths in stems of reeds) |
| | Alder flies |
| | Caddis flies |

chalk river in southern England, mute swans significantly reduced *Ranunculus* biomass, reducing the ability of the river to deliver ecosystem services such as biomass production for fish. Indeed, as well as eating plants, each swan on the river was, when eating plants, incidentally consuming as many invertebrates per day as a 300 g trout (O'Hare et al. 2007).

Macrophytes are also important oviposition sites for 2 main groups of animals: invertebrates and fish. Among the macroinvertebrates, a wide cross section of species use macrophytes for this purpose (Table 6.1). Of the European fish fauna, about 20% are exclusively phytophilous in egg laying, with a further 15% to 20% sometimes using plants.

### 6.1.2.4  Role of Macrophytes in Supporting Ecosystem Functions

Macrophytes have a variety of roles in maintaining the physical and chemical functioning of freshwater ecosystems. Plants generally increase water clarity, probably through the combination of a variety of factors, including reducing sediment resuspension (mainly in larger water bodies where wind action can be important), taking up nutrients, competing with algae for light, and possibly also by secreting chemicals that suppress other plants (allelopathy).

Plants also influence the dissolved oxygen climate. However, this role is not simply one of adding oxygen to the water but one in which plants both add and remove oxygen from the water. For example, in smaller waters typical of agricultural landscapes, plants often create a considerable oxygen swing, in some cases reducing the dissolved oxygen levels compared to open water. For example, in an upland pond in Wales, where dissolved oxygen concentrations in the open water were stable, vegetation caused a diurnal drop in dissolved oxygen concentrations (Figure 6.4).

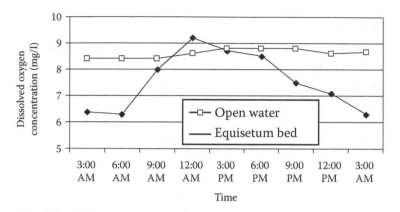

**FIGURE 6.4**   Diurnal variation in dissolved oxygen concentrations in a pond in mid Wales in open water and amongst vegetation (Laurie 1948; reprinted with permission from Laurie EMO. The dissolved oxygen of an upland pond and its inflowing stream, at Ystumtuen, North Cardiganshire, Wales. *J Ecol* 30: 357–382. Wiley-Blackwell.)

### 6.1.3   PATTERNS OF MACROPHYTE BIODIVERSITY IN THE AGRICULTURAL LANDSCAPE

Comparative data on the relative species richness of aquatic plant assemblages in ditches, streams, ponds, lakes, and rivers are available from landscape studies by Williams et al. (2004) and Davies et al. (2008). These studies describe the patterns seen in species richness in aquatic macrophytes (and also invertebrates) across some representative lowland agricultural landscapes. The most detailed data are available from the UK where both plant and invertebrate data are available; only plant data are available from the continental European study sites. The studies are important because they represent one of the few attempts to describe aquatic biodiversity in a range of waterbody types across the landscape, in contrast to most traditional approaches, which focus on one type of waterbody. We are not aware of any previous studies where it has been possible to make strictly comparable (i.e., based on the same sampling methodology) assessments of the relative contribution to biodiversity of different waterbody types in this way. These studies also collected data on plant abundance, which have not previously been published, some of which are presented below.

The study focuses on the analysis of alpha and gamma diversity. Alpha diversity is the number of species seen at a site — the number of plant species in a 500-m length, the number of invertebrate species in a hand net or surber sample. Gamma, or regional, diversity is the total number of species found in an area. It is particularly important in studies of freshwater because many species are mobile and make use of a range of habitats in a landscape, and it is arguably a more realistic representation of the freshwater biodiversity of a region. As long as the region retains its species pool, landscape management can be regarded as successful. Note that these results are for vascular plants only and exclude bryophytes, algae and charophytes.

In the UK, in a central southern England study area (Coleshill), macrophyte site-level (alpha) species richness was greatest in rivers and ponds, and lowest in ditches (mean number of wetland plant species per 75 m² survey area: rivers 10.7, ponds 10.1, streams 7.3, ditches 6.1). In a second UK site (Whitchurch), an area that lacked rivers, ponds were the richest water bodies, followed by ditches and streams (Williams et al. 2004).

In contrast, in both Coleshill and Whitchurch, regional (gamma) diversity was greatest in ponds. The ponds also supported the greatest number of uncommon species. Rivers, streams, and ditches supported fewer plant species regionally. Similar patterns were seen in 3 other areas in continental Europe: Funen (Denmark), Braunschweig (Germany), and Avignon (France) (Figure 6.5; Davies et al. 2008). In all of these areas, the largest number of species was contributed by ponds, although in the Avignon areas, the differences between waterbody types were relatively small.

The macrophyte flora of water bodies in agricultural landscapes is typically dominated, in terms of species richness and percentage cover, by emergent plants, with smaller numbers of submerged and floating-leaved species (Figure 6.6). Ditches often completely lack submerged species. For example, ditches in the Braunschweig area lacked submerged plants and in the Coleshill area lacked both submerged and aquatic species. Generally, species richness and abundance of submerged and floating-leaved plants is low, with typically less than 2 species per site, and cover less than 10% (Figure 6.6).

It is often assumed that *Lemna* is frequent and abundant in all landscape types and all standing waterbody types, including ditches. Distribution maps for the species, such as those for the United Kingdom and Germany (see, respectively, the map

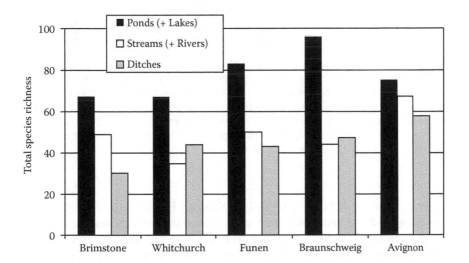

**FIGURE 6.5** Regional macrophyte (gamma) diversity in rivers, ponds, streams, and ditches in the agricultural landscape of Europe. (Reprinted from *Agric Ecosyst Environ*, 125, Davies B, Biggs J, Williams P, Whitfield M, Nicolet P, Sear D, Bray S, and Maund S, pp.1–8, © 2008, with permission from Elsevier.)

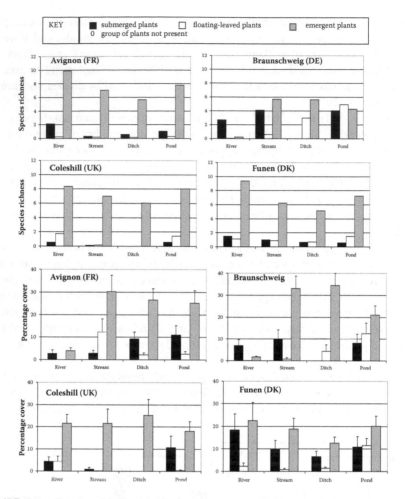

**FIGURE 6.6**  Species richness and percentage cover of submerged, floating-leaved and emergent macrophyte in water bodies in agricultural landscapes in 4 representative European regions: around Avignon (France), Braunschweig (Germany), Coleshill (UK), and Funen (Denmark). In each study, the sample size was between $n = 80$ (Coleshill) and $n = 92$ (Avignon). Details of the study areas for these previously unpublished data are contained in Williams et al. (2004) and Davies et al. (2008).

of *Lemna minor* distribution on the UK National Biodiversity Network at http://data.nbn.org.uk/ and the distribution of *L. minor* in Germany on the Bundesamt für Naturschutz site at http://www.floraweb.de/index), reinforce the impression that *Lemna* is common everywhere. However such maps hide the local pattern of *Lemna* distribution. In low-relief landscapes, such as those of Europe's drained wetlands and river valleys, it may be common in all kinds of waterbody. Elsewhere it is often less frequent than imagined at the landscape level.

For example, in the Coleshill area of southern England, which is dominated by two of the most widespread landscape types in Britain (Brown et al. 2006), 3 *Lemna* species were present (*L. minor, L. minuta,* and *L. trisulca*) but all were comparatively infrequent, except in ponds. Overall, *Lemna* species were found in 14% of a stratified random sample of rivers, streams, ditches, and ponds. *Lemna* species were most frequent in ponds but were not recorded at all in ditches, mainly because many were seasonal, and were present in only 5% of streams and 15% of river sites. *Lemna* cover in all habitats was low, with overall mean cover of only 1.2%.

Similarly, in the intensively agricultural area of Braunschweig (Germany), *Lemna minor* was the most widespread species but was present in only 30% of water bodies. Within the sites where it occurred, the cover of *L. minor* averaged 30.4%. In the agricultural landscape of Funen (Denmark), *L. minor* was more widespread, being found in 65% of water bodies but with cover, averaging 7.5%, generally low (Pond Conservation, unpublished data, with permission). In Dutch ditch networks, a similar situation occurs: *Lemna gibba* and *L. minor* occurred respectively in 68% of vegetation-rich, eutrophic ditches (Nijboer et al. 2004).

Like *Lemna*, other aquatic plants also have patchy distributions but often have higher cover values (i.e., are more abundant where they occur). Thus in Coleshill, where 22 submerged aquatic species were recorded, 82% were more abundant where they occurred than *Lemna*. For example, *Myriophyllum spicatum*, which occurred in 11% of sites, had a mean percentage cover value of 4%. *Elodea nuttallii*, which occurred in only 4% of sites, had a mean cover value of 13%. Similarly, in Braunschweig where *Lemna* was the most widely occurring aquatic plant, about one-third of submerged plant species were more abundant where they occurred than *Lemna* (Table 6.2). In Dutch ditch networks, the submerged macrophyte *Elodea nuttallii* is recorded in 50% of vegetation-rich, eutrophic ditches (Nijboer et al. 2004).

### 6.1.4  CONCLUSIONS

Macrophytes are a significant component of freshwater biodiversity with important biodiversity and functional roles.

- There is clear evidence of impairment of wetland flora in farmland, although the relative importance of different causes (sediments, nutrients, chemical pollutants) is unclear.
- The vascular wetland flora in agricultural landscape is dominated by emergent species, in terms of species richness and percentage cover.
- In drained wetland landscapes, such as those of the Netherlands, *Lemna* species can be very frequent. For example *L. gibba* and *L. minor* occurred in 68% of vegetation-rich, eutrophic ditches.
- Ditches outside of drained wetlands may lack submerged aquatic species completely. Submerged and floating-leaved plant species richness is generally low, with cover typically below 10% in all waterbody types in agricultural landscapes.

## TABLE 6.2
## Occurrence and percentage cover of *Lemna minor* in 2 representative European agricultural landscapes: Coleshill (UK) and Braunschweig (DE)

| Species | Frequency | | Abundance (% Cover) | |
|---|---|---|---|---|
| | Coleshill | Braunschweig | Coleshill | Braunschweig |
| **Submerged Plants** | | | | |
| *Callitriche obtusangula* | 5% | – | 14.3 | – |
| *Callitriche stagnalis/platycarpa agg.* | 16% | 10% | 0.2 | 12.5 |
| *Callitriche hamulata/brutia agg.* | 3% | 1% | 6.1 | 28.5 |
| *Ceratophyllum demersum* | 4% | 3% | 0.4 | 86.6 |
| *Ceratophyllum submersum* | – | 1% | – | 96.8 |
| *Elodea canadensis* | 1% | 1% | 0.1 | 0.1 |
| *Elodea nuttallii* | 4% | 2% | 13.4 | 80.2 |
| *Hippuris vulgaris* | – | 1% | – | Not present in quadrats |
| *Hottonia palustris* | – | 1% | – | 0.1 |
| *Juncus bulbosus* | – | 1% | – | 17.0 |
| *Lagarosiphon major* | 3% | – | 41.5 | – |
| *Myriophyllum aquaticum* | 1% | – | 0.1 | – |
| *Myriophyllum spicatum* | 11% | 1% | 4.0 | 10.0 |
| *Potamogeton berchtoldii* | 3% | 1% | 2.3 | 0.5 |
| *Potamogeton crispus* | 3% | 1% | 1.1 | 5.0 |
| *Potamogeton pectinatus* | 4% | 9% | 1.0 | 32.0 |
| *Potamogeton pusillus* | 4% | 1% | 0.7 | 3.0 |
| *Ranunculus aquatilis* | 1% | – | 8.0 | – |
| *Ranunculus penicillatus* | 5% | – | 3.3 | – |
| *Ranunculus trichophyllus* | 3% | 1% | 4.0 | 48.0 |
| *Ranunculus* sp. (undet.) | 3% | 1% | 0.6 | 2.0 |
| *Sagittaria sagittifolia* | 1% | – | 0.1 | – |
| *Sparganium emersum* | 4% | 2% | 3.4 | 14.2 |
| *Stratiotes aloides* | – | 1% | – | 36.7 |
| *Zannichellia palustris* | 3% | – | 12.6 | – |
| *Fontinalis antipyretica* | 4% | – | 0.7 | – |
| *Chara* sp. | 1% | – | 2.0 | – |
| **Floating-Leaved Plants** | | | | |
| *Hydrocharis morsus-ranae* | – | 2% | – | 19.0 |
| *Lemna gibba* | – | 1% | – | 64.5 |
| *Lemna minor* | 14% | 31% | 0.3 | 30.4 |
| *Lemna minuta* | 6% | – | 3.5 | – |
| *Lemna trisulca* | 3% | 2% | 0.1 | 24.2 |
| *Nuphar lutea* | 9% | 1% | 12.7 | Not present in quadrats |
| *Nymphaea alba* | 1% | – | 0.1 | – |

*(Continued)*

**TABLE 6.2    (Continued)**

| Species | Frequency | | Abundance (% Cover) | |
|---|---|---|---|---|
| | Coleshill | Braunschweig | Coleshill | Braunschweig |
| **Floating-Leaved Plants** | | | | |
| *Nymphaea* sp. (ornamental) | – | 1% | – | Not present in quadrats |
| *Persicaria amphibia* | – | 9% | – | 6.6 |
| *Potamogeton natans* | – | 4% | – | 23.3 |
| *Riccia fluitans* | – | 2% | – | Not present in quadrats |
| *Spirodela polyrhiza* | – | 4% | – | 25.5 |

## 6.2    REGULATORY ISSUES WITH RESPECT TO THE RISK ASSESSMENT OF MACROPHYTES

Peter van Vliet (CTGB, NL), Anne Alix and Véronique Poulsen (AFFSA, FR),
Paul Ashby (PSD, UK), Martin Streloke and Petra Pucelik-Günther
(Bundesamut für Verbraucherschutz und Lebensmittelsicherheit, DE)

### 6.2.1    INTRODUCTION

Under existing risk assessment procedures in the EU, the risk of herbicides to aquatic plants and algae is initially evaluated by calculation of toxicity exposure ratios (TERs) between toxicity endpoints (EC50) derived from standard laboratory tests with recommended test species and predicted environmental concentrations (PECs). The resulting TER is compared with the trigger of 10, defined in Annex VI of 91/414EEC (EU 1997). TER values that exceed this trigger indicate that the compound under evaluation can be considered not to pose an unacceptable risk to aquatic plants and algae, whereas TER values that fall below this trigger indicate a potential unacceptable risk and indicate the need for a higher-tier risk assessment. The higher-tier assessment may involve refinement of exposure values, often through mitigation measures including drift reduction techniques and buffer zones, or generation of further toxicity endpoints for additional species, possibly under field conditions, in order to reduce uncertainty in the risk assessment.

The risk assessment for aquatic macrophytes is currently based on the recommended test species, *Lemna*. However, uncertainty has been expressed about the suitability of *Lemna* as an indicator species for aquatic macrophytes. In particular, concern has been expressed that *Lemna*, being a non-rooted, monocot may not be sensitive to residues in sediment or modes of action unique to dicot species. As a result of these concerns, a number of regulatory issues relating to the aquatic macrophyte risk assessment are raised as a series of questions. These issues were intended to promote discussion during the workshop, and where available, information that could be useful in addressing them is given in italics.

The regulatory issues are grouped under the following headings:

- first-tier issues
- higher-tier issues
- recovery issues
- (semi-field) tests
- ecological modeling
- linking fate and effects

## 6.2.2  FIRST TIER

For some modes of action and for some exposure routes, concerns have been raised over the existing risk assessment scheme. Examples include the situation when the dominant exposure route is via the sediment. The use of *Lemna* will not address this exposure route.

- Only test protocols for *Lemna* and *Myriophyllum* are available at the moment (OECD 2006c; ASTM 2007). Only the test protocol for *Lemna* is internationally accepted. The UK has a national protocol for *Glyceria maxima*, a rooted monocot species (Davies 2001). This protocol has not been ring tested.
- This issue is elaborated in Chapter 3 and is being progressed by AMRAP Workgroup 2 (Chapter 5).

The selection of relevant endpoints is critical in order to accurately assess effects and to use in risk assessment. There is no current agreement on what the endpoints to assess effects on aquatic macrophytes should be.

- From a paper by Arts et al. 2008, it was concluded that evaluation criteria for macrophyte toxicity tests should comprise not only less sensitive endpoints like growth or biomass and shoot endpoints, but an array of endpoints including very sensitive endpoints like root growth. Most rooted macrophytes can be evaluated for a wide variety of endpoints, which may be more representative of macrophyte fitness than biomass and growth only.

In the current EU risk assessment process, the EC50 is taken as the relevant endpoint for algae and aquatic plants. It may be more logical to take the EC10 or NOEC as the relevant endpoint.

- The test for *Lemna* covers both acute and chronic effects due to the rapid vegetative reproduction, while other aquatic macrophytes reproduce much more slowly. The proposal in the revision of Directive 91/414/EEC (Annex II and III) is to take the NOEC for the current *Lemna* test as the endpoint as opposed to the EC50. With the same TER of 10 for algae and *Lemna*, this approach will make the risk assessment for algae and macrophytes more conservative.

A critical issue is the question of when *Lemna* is or is not appropriate for assessment of certain active substances. It is known that *Lemna* is insensitive to, for example, auxin simulators, which are more toxic to submerged plants. Hence there is a necessity to address the issue of an alternative additional species at Tier 1.

- There is some information available from an article by Vervliet-Scheebaum et al. 2006: comparing plant species across the different experiments, the range of sensitivities identified show that no single plant species is always the most sensitive, even for compounds with the same mode of action, but it should be noted that differences in the testing method (e.g., emerged, submersed, or rooted form of the plant in the test, temperature, test medium, pH, light intensity) or changes in the application method can lead to substantial differences in the values recorded as test endpoints.
- Results presented by Brock et al. (2000) showed that for more than 80% of the compounds — mainly photosynthesis-inhibitors — the existing testing scheme with a green algae and *Lemna* sp. was sufficient to detect potential toxicity against non-target aquatic plants. But the results also indicate that further test species need to be identified for testing the impact of auxin-simulating herbicides or grass-specific compounds to evaluate if toxicity to non-target aquatic plants is underestimated.
- The issue was a critical area of debate in the workshop and is covered extensively in Chapters 2 and 3. It is also being followed up by AMRAP Workgroup 1.

It is unclear what drives the differences in sensitivity between dicotyledons and monocotyledons and whether this is solely dependent upon the MoA of the active substance. Evidence is that MoA is not the only criterion.

## 6.2.3 Higher Tier

It is clear that additional information is needed for higher-tier assessments using aquatic macrophytes in terms of both the tools needed for assessment, and guidance as to their use and interpretation. These include modified exposure tests, recovery tests, SSD approaches, semi-field and field tests, and ecological modeling.

- These issues were debated during the workshop because they arose from the AMRAP case studies and from general discussion. In particular, the SSD approach is detailed in Section 3.2.4 and is taken up by AMRAP Workgroup 4.

There are no formalized test guidelines available or under development for modified exposure tests (water or sediment), recovery tests, or (semi-)field tests with macrophytes and it is questionable whether this is feasible at higher tiers. According to HARAP (Campbell et al. 1999), higher-tier studies should be designed to answer the questions posed; hence standardization was considered inappropriate. However, some guidance is needed regarding the acceptability criteria for certain approaches, for example, the number of species to be used in an SSD and what the endpoints should be based upon.

- HARAP guidance states that a minimum of 8 species is needed to construct an SSD.

Guidance is also required concerning which endpoints should be taken into account for the determination of the acute HC5 value and the chronic HC5 value (NOEC or EC10 or EC50) and whether the median HC5 or the lower-limit HC5 is an appropriate endpoint for risk assessment.

- In the paper by Van den Brink et al. (2006), the following useful information is stated with respect to this issue: "Hazardous concentrations estimated using laboratory-derived acute and chronic toxicity data for sensitive freshwater primary producers were compared to the response of herbicide-stressed freshwater ecosystems using a similar exposure regime. The lower limit of the acute $HC_5$ and the median value of the chronic $HC_5$ were protective of adverse effects in aquatic microcosms and mesocosms even under a long-term exposure regime. The median $HC_5$ estimate based on acute data was protective of adverse ecological effects in freshwater ecosystems when a pulsed or short-term exposure regime was used in the microcosm and mesocosm experiments."

The application of an appropriate assessment factor in higher-tier acute and chronic risk assessment of aquatic macrophytes is part of an ongoing debate. Little guidance is available, and experience with mesocosm data, mostly focusing on effects on invertebrates, indicates a diversity of regulatory opinion regarding what assessment factors should apply to the selected endpoints.

### 6.2.4   RECOVERY ISSUES

It may be possible to measure recovery potential following an effect such as growth inhibition, but guidance is needed on how to incorporate into the risk assessment. Tools are also needed to be able to extrapolate the recovery potential for one or a few tested species to other species.

- Competition between different species must also be taken into account; if growth inhibition by a certain active compound differs between different macrophyte species, a less sensitive species may out-compete the more sensitive species (particularly if the less sensitive species also has a faster growth rate).
- Performing a multispecies toxicity test with several macrophytes (e.g., an outdoor microcosm test) may shed light on the impact of realistic exposure concentrations on these species, including recovery and indirect effects.

- Population-growth modeling, including several alternative scenarios, is another option. For example, for national registration purposes in The Netherlands, a population-growth modeling study is available with the following scenarios:
    - species that is 3× more sensitive with a 3× slower growth rate than the test species,
    - species that is 5× more sensitive with a 5× slower growth rate than the test species, and
    - species that is 10× more sensitive with a 10× slower growth rate than the test species.

## 6.2.5 Field and Semi-Field Tests

To assess effects on several macrophyte species, microcosm and mesocosm studies may be appropriate. The design of a study can influence the response of aquatic macrophytes to a chemical. In some cases it may be best to perform bioassays on selected species, while in others a mesocosm using naturally established macrophytes may be more appropriate.

- There is some information available from an article by Coors et al. (2006) that macrophytes in mesocosms may pose problems. Often, macrophyte growth and species composition are not controlled, and species developing from the sediment are tolerated without further management. This results in high variability between and within treatments and thus decreases the overall statistical power of the test system. Moreover, mesocosms are often dominated by few or even only one macrophyte species and therefore do not allow the estimation of toxicity over a range of macrophyte species.
- A possible solution is the approach of macrophyte in-situ bioassays, which have the following advantages:
    - simultaneous acquisition of toxicity data for several species of aquatic plants under more realistic conditions compared to laboratory tests, and
    - inclusion of macrophytes as important structural and functional components in mesocosms while limiting their domination of the model ecosystem.
- The above issues were debated in the workshop, supported by case studies. The participants' opinion was that, where specific species sensitivity data are required, a bioassay approach is appropriate, whereas if the study is to investigate effects in a natural system, then a naturally established mesocosm approach is appropriate (see Section 2.3.4).

## 6.2.6 Ecological Modeling

The use of ecological modeling in ecological risk assessment is discussed in the ELINK document (Brock et al. in press). There is no reason that such an approach could not be used for risk assessment of macrophytes, but further information on the spatio-temporal distribution of species is needed to allow for extrapolation of existing data sets to simulate both effects and recovery potential.

- Under "recovery issues," an example is presented of a population growth model that can add something to help address the problem of the extrapolation of the recovery potential for one or a few tested species to other species.
- Ecological modeling may be useful also to simulate whether or not recovery might occur under expected PEC conditions.

### 6.2.7 LINKING FATE AND EFFECTS

There is a need to more closely link fate and behavior characteristics to the expression of effects. In a general way this has been the subject of the ELINK workshop (Brock et al., in press). The behavior of the pesticide in natural waters can have profound consequences for exposure and effects. The rate of dissipation and the distribution between water and sediment need to be considered in the context of risk assessment, that is, whether or not the use of a PEC initial or a time-weighted-average (TWA) concentration is appropriate.

Information is also necessary to establish time to onset of effects and to inform about delayed effects.

- The use of time-weighted-average predicted environmental concentrations in surface water may be appropriate in calculating toxicity exposure ratios if the design of the effects study broadly covers the type of exposure profile indicated by the use pattern of the pesticide and its fate and behavior characteristics. This is described in Section 2.3.1 and more fully in ELINK (Brock et al., in press).

## 6.3 CRITICAL EVALUATION OF LABORATORY TOXICITY TESTING METHODS WITH AQUATIC MACROPHYTES

Nina Cedergreen (University of Copenhagen, DK), Gertie Arts (Alterra, NL), Jo Davies (Syngenta, UK), Katja Knauer (Federal Office for Agriculture, formerly University of Basel, CH)

### 6.3.1 INTRODUCTION

In the risk assessment of pesticides in the EU, *Lemna* sp. is the only macrophyte species required to be tested. The question concerning the adequacy of *Lemna* sp. as the representative for aquatic macrophytes is therefore often asked. Also, the extent to which laboratory results with *Lemna* sp. can be extrapolated to macrophyte communities in the field is often discussed. In that context, this keynote presentation evaluates the current state of the science of aquatic macrophyte testing in single-species laboratory studies.

### 6.3.2   LEMNA SP. AS A TEST SPECIES

The family Lemnaceae contains the world's smallest flowering plants. The two most frequently tested *Lemna* species are the floating species *Lemna minor* and *Lemna gibba*. *Lemna* has several advantages as a test species. It is small and therefore takes up little space in the laboratory. It grows rapidly, which means that toxicity can be detected within a relatively short time. It is clonal, allowing genetically identical plants to be used in a test. It can also be grown aseptically, thereby avoiding microbial and algal contamination of the test solutions. Its simple morphology, exposing all leaf area in one dimension, allows easy measurement of growth from the leaf surface area. The plant is floating, and therefore only the lower side of the leaves is exposed to the test media. Consequently, concern has been expressed that *Lemna* may be less sensitive to herbicides applied via the growth medium than are fully submerged species (Brock et al. 2000), which have their entire surface area exposed to the toxicant in the media. Another concern is that, because *Lemna* is not rooted in sediment, the effects of polluted sediments on macrophyte growth cannot be monitored using the standard *Lemna* test.

### 6.3.3   ROOTED MACROPHYTE TESTS AND TEST REQUIREMENTS

A large range of submerged and emergent macrophytes has been tested under laboratory conditions. However, because internationally accepted, standardized, ring-tested methods for macrophytes other than *Lemna* species do not exist, the results can be difficult to compare. Compared to *Lemna*, most other macrophyte species are larger and have slower growth rates. These features necessitate the use of larger containers and increased test durations, often from 1 to 4 weeks, though few long-term studies have been conducted in laboratory systems to evaluate toxicity and recovery over time. For recovery studies, it is especially important to include slower-growing plants than *Lemna* (and algae), because slow-growing plants are likely to recover at a slower rate than more rapidly growing plants. The largest difference from the *Lemna* test is, however, the presence of sediment and lack of sterility.

A test method using aseptic *Myriophyllum spicatum* without sediment has been proposed, but optimal growth rates cannot be achieved without the addition of sucrose to the medium, which might be problematic (Roshon et al. 1996). The presence of sediment and/or nutrients in the media often creates problems with microbial and algal growth in the media and on the macrophytes. Microbes and algae degrade chemicals and interfere with macrophyte growth through competition for light, carbon, and mineral nutrients. Hence, minimizing microbial and algal growth in macrophyte test systems is highly desired and can be partially achieved by physically separating the root and shoot media. Macrophytes can sustain maximal growth by taking up nitrogen and phosphorous from the sediment (Barko and Smart 1981a). Hence, these macro-nutrients can be omitted from the shoot media, thus reducing algal growth. Several different types of root media have been tested, ranging from agar, sand with nutrient solution, and natural soils or sediments. Good results have been achieved with specific types of standard OECD soils. Ultimately, the choice of root media depends on the properties of the test substance and the specific issue under investigation. In all cases, optimal growth of the control plants is imperative.

Optimal growth conditions in terms of irradiance for submerged plants are 100 to 200 $\mu$mol $\cdot$ m$^{-1}$ $\cdot$ s$^{-1}$ (PAR), with the *Myriophyllum* species being particularly light demanding (Madsen et al. 1991), while emergent plants often need close to 400 $\mu$mol m$^{-1}$ $\cdot$ s$^{-1}$ (PAR) to saturate photosynthesis. Optimal temperatures vary according to species, but approximately 20 °C will suffice for most temperate species. *M. spicatum* growth, which is the test species in the ASTM guideline, seems to be insensitive to temperature changes in the range of 16 to 31 °C (Barko and Smart 1981b).

### 6.3.4   ASSESSMENT PARAMETERS AND ENDPOINTS

In contrast to *Lemna*, with its simple morphology, both submerged and emergent macrophytes have very complex morphologies and growth patterns. These features make it more difficult to choose which endpoint to measure. The more commonly measured parameters include shoot length, weight, and number, and root length, weight, and number. In addition, pigment content, primarily chlorophyll and carotenoids, is often measured. Also photosynthesis, mainly as chlorophyll fluorescence, can be measured.

### 6.3.5   IS *LEMNA* REPRESENTATIVE? AN OVERVIEW OF SSDS

So, is the sensitivity of *Lemna* sp. to herbicides representative of other macrophyte and algal species? To investigate this point, we collected 16 SSDs for 13 herbicides. The majority of compounds were photosystem II inhibitors (Table 6.3). For each SSD, the position of *L. minor* and *L. gibba* in the distribution was expressed in terms of a percentile. For example, a percentile of 0.20 indicates that 20% of the species are expected to be more sensitive than the *Lemna* species. The average percentile for *L. minor* was 0.45 ± 0.24, indicating that the sensitivity of *L. minor* across herbicides is close to the average of the expected log-normal distribution of species sensitivity for algae and macrophytes. In contrast, *L. gibba* is less sensitive, with an average percentile of 0.74 ± 0.15. This conclusion, however, is based on few data. More data on both species should be included to evaluate whether this is a general trend.

Assuming that *L. minor* is representative of tested macrophyte and algae species, the variance of the log-normal species distribution was evaluated by calculating the ratio between the 5% and 95% hazard concentrations. This estimate gives a measure of the concentration range over which 90% of the species are affected (at the EC50 level). With the exception of metsulfuron-methyl, the HC5-95 was generally 20-fold or less. For 8 herbicides, the ratio was below 10, and for the 5 herbicides where the ratio was higher, *L. minor* was among the most sensitive species. In the case of metsulfuron-methyl, the SSDs were based on changes in specific leaf area because the experiments were not run for sufficient time to induce growth changes. Therefore, these SSDs cannot be directly compared to the other growth- or biomass-based SSDs. Given that robust laboratory SSDs have been shown to be representative of outdoor, long-term microcosm experiments (Van den Brink et al. 2006), more

## TABLE 6.3

**Different parameters from SSDs based on EC50 laboratory values (not including sediment) for 13 herbicides representing photosystem II inhibitors (PSII inhib.), a photosystem I electron diverter (PSI e. diverter), a microtubule assembly inhibitor (Micro. inhib.), a synthetic auxin (Synth. auxin) and an acetolactate synthase inhibitor (ALS inhib.).**

| Pesticide | Mode of Action | Endpoint | Group of Species | n | Lemna minor Perc. | Lemna gibba Perc. | HC$_{50}$ (µg/L) | HC$_5$ (µg/L) | Fold Change HC$_{5-95}$ | Reference |
|---|---|---|---|---|---|---|---|---|---|---|
| Terbuthylazine, LL | PSII inhib. | B | M | 6 | 0.29 | | 100 | 11 | 18 | Cedergreen et al., 2004 |
| Terbuthylazine, HL | PSII inhib. | B | M | 9 | 0.40 | | 158 | 39 | 8 | Cedergreen et al., 2004 |
| Atrazine | PSII inhib. | B&G | M&A | 29 | 0.47 | | 137 | 13 | 21 | Van den Brink et al., 2006 |
| Atrazine | PSII inhib. | G | M&A | 25 | 0.40 | | No fit | No fit | <10 | Kuster & Altenburger, 2007 |
| Metrabuzine | PSII inhib. | B&G | M&A | 19 | 0.65 | 0.95 | 29 | 7.4 | 8 | Van den Brink et al., 2006 |
| Linuron | PSII inhib. | B&G | M&A | 8 | 0.11 | | 64 | 5.8 | 22 | Van den Brink et al., 2006 |
| Simazine | PSII inhib. | B&G | M&A | 10 | 0.45 | 0.73 | 221 | 52 | 9 | Van den Brink et al., 2006 |
| Metamitron | PSII inhib. | B&G | M&A | 4 | 0.80 | 0.60 | 1214 | 667 | 4 | Van den Brink et al., 2006 |

*(Continued)*

**TABLE 6.3 (Continued)**

| Pesticide | Mode of Action | Endpoint | Group of Species | n | Lemna minor Perc. | Lemna gibba Perc. | HC$_{50}$ (µg/L) | HC$_5$ (µg/L) | Fold Change HC$_{5-95}$ | Reference |
|---|---|---|---|---|---|---|---|---|---|---|
| Diuron | PSII inhib. | B&G | M&A | 7 | 0.63 | | 21 | 12 | 4 | Van den Brink et al., 2006 |
| Isoproturon | PSII inhib. | G | A | 12 | 0.92 | | No fit | No fit | <10 | Kuster & Altenburger, 2007 |
| Prometryn | PSII inhib. | G | M&A | 10 | 0.55 | | No fit | No fit | <10 | Kuster & Altenburger, 2007 |
| Diquat | PSI e.diverter | B&G | M&A | 15 | 0.13 | | 34 | 3.5 | 19 | Van den Brink et al., 2006 |
| Pendimethalin | Micro. inhib. | B&G | M&A | 5 | | 0.67 | 11 | 2 | 11 | Van den Brink et al., 2006 |
| 2,4-D | Synth. auxin | B&G | M&A | 11 | 0.50* | | 558 | 71 | 16 | Van den Brink et al., 2006 |
| Metsulfuron, LL | ALS inhib. | SLA | M | 8 | 0.22 | | 0.79 | 0.031 | 51 | Cedergreen et al., 2004 |
| Metsulfuron, HL | ALS inhib. | SLA | M | 5 | 0.17 | | 0.4 | 0.014 | 57 | Cedergreen et al., 2004 |

* Estimated from EC50 values from MCPA and mecoprop of approximately 10 000 µg/L

information on macrophyte SSDs could potentially improve aquatic macrophyte risk assessment.

For terbuthylazine and metsulfuron-methyl, the SSDs were based on data for plants grown under low light (LL) and high light (HL) conditions. The SSD endpoints are both biomass at harvest (B) or relative growth rates of macrophytes (G) or of algae (A), which are included in the SSDs where algae SSDs were not different from macrophyte (M) SSDs. For one herbicide, specific leaf area (SLA) was the only sensitive endpoint across all the tested species. The percentiles for *L. minor* and *L. gibba* are given. For the curves that were described with a symmetric sigmoid curve (log-normal), the ratio between the 5% hazard concentration (HC5) and the 95% hazard concentration (HC95) is also given.

### 6.3.6   WHICH ENDPOINT IS MORE SENSITIVE? DEPENDENCE ON PESTICIDE MODE OF ACTION

The diverse morphology of macrophytes enables measurement of a wide range of morphological endpoints, including side shoot number, main- and side-shoot length and shoot biomass, root number, length, and biomass. In addition, physiological endpoints such as pigment content are also commonly measured. The relative sensitivity of these endpoints and the potential relationship between endpoint sensitivity and mode of action has been investigated. A review of endpoint sensitivity for 22 toxicants representing 7 known herbicide modes of action, and some unknowns, did not show any distinct pattern in sensitivity. The only exception was the synthetic auxins, where pigment content appeared to be the most sensitive endpoint. Also, effects on photosynthesis have proved to be a quick and sensitive endpoint for photosynthetic inhibitors, while less so for toxicants with other modes of action. It is therefore recommended that a wide range of endpoints be monitored, to get the best picture of the overall toxicity. When the number of measurements have to be limited to a few, biomass-related endpoints are to be preferred, as those are expected to be the ecologically most relevant endpoint. Other physiological and morphological endpoints can be informative in regard to identifying modes of action of unknown chemicals, and, if non-destructive, to determine the kinetics of the toxicity and recovery.

### 6.3.7   CONCLUSION

The *Lemna* test has many advantages, and compared to other macrophytes, *Lemna minor* seems to be representative. However, several ecologically relevant issues cannot be investigated using only *Lemna*. In particular, concern has been expressed that *Lemna* may be less sensitive to herbicides applied via the growth medium than are fully submerged species, and that, because *Lemna* is not rooted in sediment, the negative or positive effects of sediment on toxicity cannot be monitored. In addition, the rapid growth rates seen in *Lemna* species may lead to toxicokinetic and recovery differences relative to slower-growing species. Also, in order to generate representative SSDs for aquatic primary producers, a good representation of a diverse group of macrophytes is needed.

There are several challenges in macrophyte studies. These include maintaining optimal growth rates while minimizing microbial and algal contamination for the submerged species. Also, reducing variability of the obtained data is a major challenge, which might be remedied to some extent by introducing standardized test procedures. These test procedures should include information on plant source, physical setup, nutrient media and sediment composition, irradiance and temperature levels, minimum number of doses and replicates, test duration and criteria for minimum growth of controls, along with the other quality criteria associated with standardized test protocols. Also, a wider range of modes of action should be tested systematically, in order to be able to identify mode of action specific sensitive endpoints, response, and recovery times.

## 6.4 CRITICAL EVALUATION OF (SEMI-)FIELD METHODS USING AQUATIC MACROPHYTES

Mark Hanson and Erin McGregor (University of Manitoba, CA),
Silvia Mohr (UBA, DE)

### 6.4.1 WHY MACROPHYTES AND MICROCOSM AND MESOCOSM TOXICITY TESTING?

Model ecosystems such as lentic microcosms and mesocosms are considered a useful intermediate between laboratory-based testing and full-scale field assessments for many organisms, including macrophytes. In contrast to the simple single-species laboratory-based toxicity testing, simulated field studies such as mesocosms can allow for testing of multiple species simultaneously and interactions across several trophic levels (Figure 6.7). These systems allow for the observation of indirect, or ecological, effects in macrophyte population and community structure that may occur due to modifications in nutrient availability, water quality, and macrophyte grazing, among other things (Solomon 1996; Caquet et al. 2000). Mesocosms may also provide more realistic exposure scenarios as compared to laboratory testing; in the mesocosms, the chemical stressors tested have the potential to partition, degrade, and dissipate as they would in the environment. Mesocosm testing also allows for replication between test units and permits researchers to capture the responses of organisms to a range of stressor concentrations, which may not be achieved in large-scale field investigations. The drawbacks to mesocosm-based assessments have been reviewed extensively (Shaw and Kennedy 1996), and include cost considerations, limitations in the number of replicates, and the variability of results, both biological and statistical, associated with duration and timing of studies and test locations.

Previously, field- and mesocosm-based assessments rarely focused directly on the impacts of environmental contaminants on the plant component of the aquatic ecosystem (Huggins et al. 1993). This approach has changed in the past decade, with more mesocosm work on the characterization of toxicity by chemical stressors to macrophytes being published in the literature (Huber 1994; Caquet et al. 2000; Hanson et al. 2001; Mohr et al. 2007). Within an EU risk assessment framework

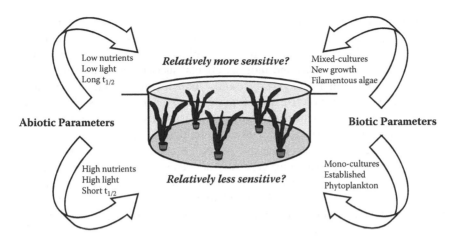

**FIGURE 6.7**  Abiotic and biotic factors can influence the relative toxicity exhibited by macrophytes in mesocosm testing. For example, where chemical half-life is short (due to, e.g., enhanced microbial degradation as a result of high nutrient levels), toxicity would be anticipated to be reduced as compared to when the half-life, and hence exposure, is extended. Biotic factors too play a role in the relative sensitivity to a stressor. For example, established populations and communities appear to be less sensitive to herbicides than newly developing macrophyte assemblages. Still, many of these relationships can interact in ways not illustrated by this figure.

context, higher-tier studies using mesocosms may be conducted if acute and chronic studies at lower tiers indicate potential risk to ecosystems of a tested substance. The OECD (2006) guideline for freshwater lentic field tests recommends the introduction of macrophytes in the mesocosms if an herbicide is being evaluated. Ideally, macrophytes would be introduced into any field system due to their obvious ecological importance and their role in stabilizing the mesocosms physically, chemically, and biologically.

## 6.4.2  What Mesocosm Macrophyte Toxicity Testing Methods Are Currently In Vogue?

As opposed to laboratory bioassays for toxicity testing, which normally have well validated and extensively standardized methodologies, mesocosm studies lack many of the basic recommendations required to completely understand the results generated (i.e., positive controls, initial species composition, study duration, nutrient levels, effect measures to monitor). This lack of standardization makes it difficult to interpret and compare results from different test systems and even within test systems for studies conducted at different times (Sanderson et al. 2008). One of the central issues is the introduction of the plants themselves into the mesocosms. Typically, there are two approaches to the placement of macrophytes: 1) the establishment of assemblages well in advance of treatment, which can represent natural communities and generally develop throughout the sediment layer of the mesocosm, or 2) the introduction of discrete individuals or small populations and communities that are kept separate from plants grown under the first approach. The first approach can be accomplished through the controlled introduction of a specific species of macrophyte (i.e., Van Wijngaarden et al. 2004, 2006) or to allow natural colonization of the sediments by macrophytes (Huggins et al. 1993), either of which allows for the integration of potential interactions and indirect effects, such as intraspecies competition between the plants, but do not generally allow for assessment of effects until the end of the study when a destructive sampling occurs. This approach is also associated with higher levels of statistical and biological variability, especially where natural colonization is used, because the plant communities between systems can be quite different. As well, there is an extensive period for colonization or establishment of the plant populations, something generally not required for bacterial, phytoplankton, or zooplankton assemblages, adding to the overall cost of these studies. This approach also means the rate at which toxicity is manifest and the possible recovery of the macrophyte community, which itself is not well defined, are generally not assessed with plants introduced in this manner.

The alternative approach, the introduction of plants as discrete individuals (Hanson et al. 2001) or as model populations and communities (McGregor et al. 2007, 2008; Rentz et al. 2009, allows for sampling at intermediate time points throughout the exposure and thus characterization of the progression of toxicity and recovery if applicable. This method demonstrates levels of variability similar to those observed with laboratory assays, and allows for the measurement of numerous endpoints as opposed to just final biomass or percent cover (Hanson et al. 2001, 2003). The disadvantage is that true interactions within and between species may not be captured and vegetation-driven sediment–water interactions may be underestimated, but it does allow for more than one species to be assessed simultaneously for a variety of responses without extensive time required for the plants to become established.

### 6.4.3 What Testing Approaches Do We Recommend with Macrophytes and Mesocosms?

Ideally, a mesocosm study design would include some combination of the two approaches to plant introductions when possible. For example, a natural population would be established and then prior to the introduction of a stressor, discrete potted individuals or model populations and communities could be introduced to allow for sampling throughout the exposure duration (UBA 2007). An approach that is integrative of the two strategies is to examine an endpoint nondestructively for the population of naturally or introduced macrophytes. This approach has been used with *Potamogeton natans*, where number and area of floating leaves may be measured using image analysis (Berghahn et al. 2006). This approach is limited to the small number of species with this morphological trait. Still, there is little information on the relative toxicological sensitivities of established macrophyte populations and newly introduced plants.

A recent mesocosm study examining the effects of the herbicide diuron on *Elodea canadensis* found that established stands (i.e., those allowed to develop from cuttings for 2 months) were less sensitive than their equivalent newly introduced plantings, especially at longer exposure durations (Rentz et al. 2009). The implication is that mesocosm studies looking only at established macrophyte stands to characterize toxicity are going to underestimate effects on newly developing plants. *E. canadensis* grown as model populations or in communities with other plants (McGregor et al. 2007; Rentz et al. 2009) were found to have higher relative growth rates (RGRs) than those grown as individuals. The RGR can have implications for the assessment of toxicity in macrophytes and other organisms. These plant species with higher RGRs have been found to be more sensitive in laboratory studies (Cedergreen et al. 2004a, 2004b) and between assays with *Lemna gibba* for the same compounds (Huebert and Shay 1993). Moreover, those test systems that use only individual plants to assess toxicity may be underestimating toxicity relative to those systems using recently planted model populations and communities. Essentially, the exposure scenario of the stressor under evaluation is important when determining the risk to macrophytes. Is exposure of the compound to established communities of macrophytes or to newly forming populations and assemblages? The mesocosm study should be designed to capture the most relevant exposure whenever possible.

While these individual planting methods are used in attempts to reduce variability within the test systems, there is no confirmation in the literature that the responses of individually grown plants characterize accurately those of more realistically grown plant populations and communities. In essence, the mesocosms become larger, more glorified laboratory tests. Based on growth rates, we should be concerned that individuals or small groupings of individuals do not adequately describe toxicity, possibly being less sensitive than model populations or communities (McGregor et al. 2007, 2008; Rentz et al. 2009).

### 6.4.4 WHAT ELSE SHOULD WE CONSIDER IN OUR MESOCOSM MACROPHYTE STUDIES?

The importance of indirect or ecological effects in the response of macrophytes or any organism cannot be overstated. Consider an ecological study with two common laboratory and mesocosm test species, *Elodea canadensis* and *Myriophyllum spicatum*. Abernethy et al. (1996) grew mono and mixed cultures of these species to investigate the response of the plants to artificially imposed stress, disturbance, and inter-specific competition. The study found that, when grown in mixed culture, *M. spicatum* proved to be the less competitive species, displaying a significant loss in biomass. Under these test conditions, a significant interspecific interaction was observed between species, implying that there is potential to modify the response of these plants to an environmental pollutant when they are grown or exposed to a stressor as a community. In another example, the development of roots is known to be influenced by the presence or absence of different species and the density of the plants (Spencer and Ksander 2005), meaning that the growth of these plants displays differential growth patterns and hence responses to a toxicant, depending on the species that find themselves interacting within a mesocosm. When indirect or ecological effects are anticipated in a mesocosm study, it is important to incorporate the measurement of species and parameters that will allow us to understand the mechanisms by which changes in the macrophyte populations manifest themselves (Figure 6.7). An example is the competition between macrophytes and fast-growing filamentous green algae that can remove nutrients from the mesocosm, thus inhibiting macrophyte development (Mohr et al. 2007) and therefore changing the sensitivity of macrophytes due to additional nutrient stress.

There are numerous species that have been tested in mesocosms, including but not limited to *Myriophyllum spicatum*, *M. sibirium*, *M. pinnatum*, *M. heterophyllum*, *M. verticillatum*, *Lemna gibba*, *L. minor*, *Elodea canadensis*, *E. nuttallii*, *Ceratophyllum demersum*, *Egeria densa*, *Najas* sp., *Potamageton crispus*, and *P. natans*. It is important to try to capture the range of physiologies (monocot and dicot) and growth forms (emergent, submerged, floating, rooted, free floating) so that as accurate as possible an understanding of the potential impacts of a stressor to the macrophyte community can be assessed. While the responses of macrophyte monocots and dicots under laboratory conditions have not been found to differ greatly when compared to the range observed in terrestrial plants (see Section 6.3, Critical Evaluation of Laboratory Toxicity Testing Methods with Aquatic Macrophytes), there are instances of plants in the same genus (*Myriophyllum*) displaying an order of magnitude difference in their response in mesocosms (Hanson et al. 2005). Therefore, it is important to include several species in any mesocosm evaluation of toxicity to ensure an understanding of the range of toxicity. In each mesocosm study, an attempt should be made to characterize the response of *Lemna* spp. when possible. This approach will allow for the direct comparison of the *Lemna* assay to the response in the field, and the subsequent comparison to other species tested, thereby possibly validating the use of the laboratory-based duckweed test for lower-tier assessment. It should be noted that *Lemna* are typically found in eutrophic systems and that most mesocosms would be considered mesotrophic at most; therefore it may not be possible in all cases to obtain sufficient growth to characterize toxicity.

A variety of endpoints in aquatic macrophytes may be monitored in order to evaluate the toxicity of an environmental contaminant. These include morphological endpoints such as length and biomass of roots and shoots, growth rates, and biochemical and physiological parameters including changes in pigment concentrations, enzyme activities, and carbon fixation. For a measurement endpoint to be of use, it must be toxicologically sensitive to the contaminant, thus allowing for calculation of effective concentrations. The response needs to be biologically relevant, where changes can be linked to sustained modifications at higher levels of biological organization. Additionally, endpoints of interest should demonstrate low variability in order to facilitate statistical discernment between treatment-related changes in the system and natural variation. Making generalizations regarding which endpoints are to be monitored is not entirely appropriate because endpoint sensitivities have been found to vary with test conditions and toxicants and between species. Instead, it has been suggested that a suite of endpoints be utilized in investigation of phytotoxicity (Davy et al. 2001). The measurement of biomass, which can be split into root and shoot biomass, should be a part of any mesocosm evaluation of macrophyte toxicity due to its statistical, toxicological, and biological relevance. Roots can also be an extremely sensitive measure of toxicity (Hanson et al. 2003) when the study is designed to capture such effects, usually through the introduction of plants as shoots with no root development prior to the introduction of the stressor. Of course, the selection of endpoints to be evaluated will be influenced by the needs of the risk assessment and the plant species chosen for the mesocosm study.

## 6.4.5  WHAT DO WE CONCLUDE ABOUT MESOCOSM TESTING AND MACROPHYTES?

In conclusion, while it is clear that there are substantial benefits to these types of (semi-)field studies, characterization of the effects of contaminants on aquatic macrophytes requires careful consideration of experimental design, species selection, endpoint selection, and incorporation of ecologically relevant exposure scenarios. The practice of using established macrophyte communities in the mesocosms would appear to be insufficient to capture the range of responses that these organisms might exhibit. We recommend a synthesis of both established communities or populations, ideally through the controlled introduction of specific plant species and the introduction of discrete individual and model populations and communities of macrophytes whenever possible. This approach will allow for regular sampling and a measure of the interactions that can occur between plants in a realistic manner. However, definitive methodology is not available for using macrophytes in plant studies. Many questions remain unanswered, including, but not limited to, the levels of nutrients that should be supplied to these systems, the definition of population and community for these organisms, and the density that is considered natural for each species under the specific mesocosm's physical and chemical range. In summary, the needs of the risk assessment should be explicitly considered when a mesocosm study is designed, if it is to properly capture the response of the macrophyte community in the field.

## 6.5  STAKEHOLDER OPINION ON CURRENT APPROACHES TO THE ASSESSMENT OF THE RISK OF PLANT PROTECTION PRODUCTS TO AQUATIC MACROPHYTES

Dave Arnold (Independent consultant, formerly CEA, UK) and
Melanie Kroos (EVONIK, Marl, Germany, formerly CEA, UK)

### 6.5.1  INTRODUCTION

Under Directive 91/414/EEC (EU 1997), the risk of an active substance to aquatic plants is evaluated in a first tier assessment comparing predicted environmental concentrations (PECs) in surface water with toxicity data from a single-species test, using the floating aquatic macrophyte *Lemna* sp. Further regulatory assessment of potential risk to aquatic macrophytes is not prescribed, and where concerns have been raised in a regulatory context, these have been addressed in a variety of ways, on a case-by-case basis.

In 2006, Cambridge Environmental Assessments were asked by the UK Pesticides Safety Directorate, now Chemicals Regulation Directorate, Health and Safety Executive, to undertake an investigation into the current practices and knowledge base regarding aquatic macrophyte studies in Europe and their use in regulatory risk assessments under the Plant Protection Products Directive 91/414/EEC (EU 1997).

A detailed questionnaire was sent to 20 EU Member States, Switzerland, the US Environmental Protection Agency (USEPA), and 18 European research institutes with experience in both conducting and interpreting aquatic ecotoxicology tests. In total, 16 Member States responded along with Switzerland and the USEPA. Twelve responses from research institutes, including some contract research organizations (CROs), were received. Responses to the questionnaires reflected that there was considerable uncertainty among Member States and the scientific community on this issue, combined with a wish to find a standardized approach. This uncertainty was also emphasized by the fact that 88% of the regulatory agencies and all research institutes expressed interest in a workshop on the subject. AMRAP was not organized as a consequence of the opinions expressed in the questionnaire; however, its objectives were to bring regulators, researchers, and industry together to debate the state of the science and to investigate possible improvements to aquatic macrophyte risk assessment for plant protection products. Hence, the workshop provided a vehicle for dissemination of the information received in the questionnaires and, as a consequence, provided background information to inform debate. Because AMRAP was a SETAC-sponsored workshop involving all stakeholders, industry views were sought through the design of a questionnaire along similar lines to that sent to Member States and researchers, modifying questions where necessary. This questionnaire was circulated to the major chemical companies and generics in 2007. Opinions from Member States are generally those of individual experts and may or may not represent a particular Member State position. Industry responses were received both from individual companies and the European Crop Protection Association.

## 6.5.2 EUROPEAN STAKEHOLDER RESPONSES

Questions and answers are briefly summarized as follows, incorporating stakeholder responses. Where different questions were asked of industry, these have been added. The "y" axis in the figures below shows the percentage distributions as a proportion of the total numbers of returned questionnaires.

*Q1: Are* Lemna *tests considered to be sufficient for Annex III risk assessment of plant protection products?*

A small proportion (24%) of the Member States and 8% of the research institutes believed that the *Lemna* test alone is sufficient in risk assessment, although an additional 41% of Member States responded that it would be case dependent (Figure 6.9). The majority of Member State and research institute respondents believe it to be an appropriate species for certain modes of action but not all. Industry responses demonstrated a preference for the use of *Lemna* only at Tier 1, although some would like to see more research to investigate what value would be added to the risk assessment by using additional species (Figure 6.8). However, higher-tier assessments may be indicated on a case-by-case basis. There are differing views about the ability of *Lemna* sp. to assess toxicity of auxin-like compounds.

*Q2: Is there a gap in the current risk assessment scheme and would additional macrophyte data improve it?*

A majority of research institutes and Member States believe that there is a gap in the current risk assessment scheme that would be improved by additional aquatic

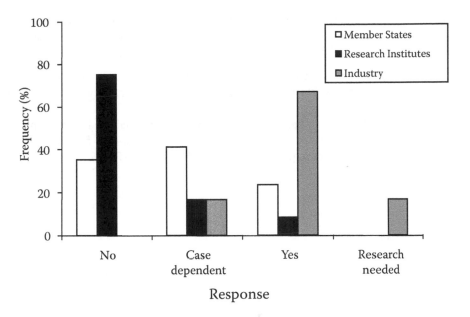

**FIGURE 6.8**   Responses of Member States, research institutes, and industry to Question 1.

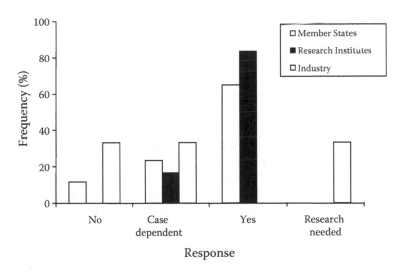

**FIGURE 6.9**   Responses of Member States, research institutes, and industry to Question 2.

macrophyte data (Figure 6.9). One third of the industry believes there is a gap, while others either see the need for more research information or see that any gap should be assessed on a case-by-case basis (Figure 6.9). Industry would like a comparative assessment of the substance toxicity to green algae and *Lemna* to be undertaken to evaluate whether taking the highest toxicity value and adding an assessment factor would provide sufficient protection for all aquatic macrophyte species. Higher-tier tests are seen as an option to reduce uncertainty beyond Tier 1.

*Q3.1: To Member States: are there instances in which industry has voluntarily (without your requesting it) submitted additional aquatic macrophyte data as part of a risk assessment?*

Instances where industry has voluntarily provided additional aquatic macrophyte data are few. The majority of Member States (88%) state that data from additional macrophyte studies are not regularly volunteered by notifiers (Figure 6.10).

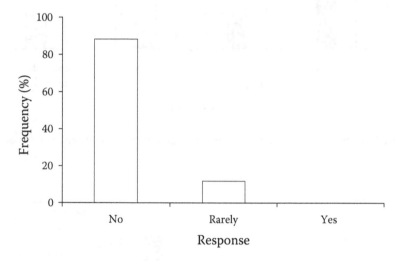

**FIGURE 6.10**  Member State responses to Question 3.1.

*Q3.2: To researchers and CROs: are you regularly requested by industry to conduct aquatic macrophyte tests on species other than* Lemna?

A small percentage (17%) of research institutes and CROs regularly conduct tests for industry on macrophytes other than *Lemna*, and about one third do so occasionally (Figure 6.11). There is no indication of how these data were used, if at all, in plant protection product risk assessment.

*Q3.3: To Industry: have you voluntarily (without it being requested) submitted additional aquatic macrophyte data to regulatory agencies to be considered in risk assessment?*

Several companies have voluntarily submitted additional aquatic macrophyte data as part of a regulatory package without it being requested by regulatory agencies (Figure 6.12). This approach was taken to inform the risk assessment. However, because these are higher-tier data, there is no guidance on either the way in which regulatory agencies should interpret such data or how it should be incorporated into risk assessment.

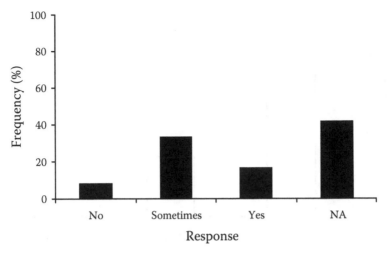

**FIGURE 6.11**   Research institute responses to Question 3.2.

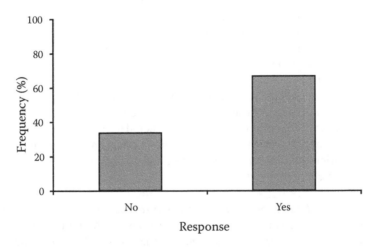

**FIGURE 6.12**   Industry responses to Question 3.3.

*Q4: To Member States: have you, for Member State evaluations, requested that notifiers provide you with additional data on macrophytes as either single-species or multi-species, for example, mesocosm, studies?*

Of the 16 Member States that responded, less than a third has requested additional aquatic macrophyte data from notifiers as part of a regulatory submission (Figure 6.13). When regulatory agencies have requested additional aquatic macrophyte data from industry, this request has been mainly in circumstances where the *Lemna* TER is <10, or there are concerns about uptake from sediment by rooted macrophytes.

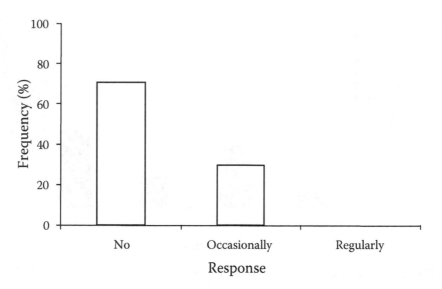

**FIGURE 6.13**    Member State responses to Question 4.

*Q5: Is additional macrophyte study data of acceptable quality?*

Only about half of the Member States surveyed have experience in evaluating additional aquatic macrophyte data. Data quality is clearly acceptable in many instances, but the general view is that there is insufficient experience among Member State experts in the evaluation of additional macrophyte species data, which makes study data-quality assessment difficult. Researchers and regulatory study practitioners remarked that different measurement endpoints have inherently differing coefficients of variance. This problem is an issue when determining the quality of data provided. For example, growth rate as a measure of effect has a higher inherent coefficient of variation than mortality, but it may, nonetheless, be the more appropriate metric for aquatic macrophytes.

*Q6: Can additional macrophyte data be used to reduce an assessment factor?*

One of the key issues is a lack of experience among Member State experts in evaluating aquatic macrophyte data, and another is a lack of guidance regarding how it should be applied in risk assessment. However, the majority (71%) of Member States would consider using additional macrophyte data in a risk assessment to reduce the assessment factor (Figure 6.14). Clearly, industry considers that additional data that helps reduce uncertainty should result in a reduced assessment factor (Figure 6.14). Some companies believe that terrestrial plant toxicity data can help in this regard.

*Q7: To Member States: have you reduced an assessment factor based upon additional aquatic macrophyte data?*

Responses in Figure 6.15 infer that only 3 Member States have practically utilized such data in risk assessment, on a case-by-case basis, leading to a reduction in the TER.

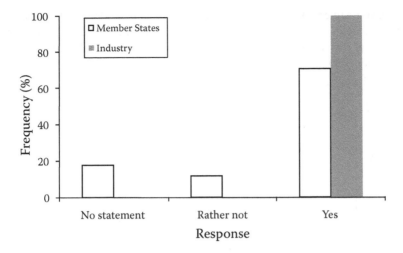

**FIGURE 6.14**   Member State and industry responses to Question 6.

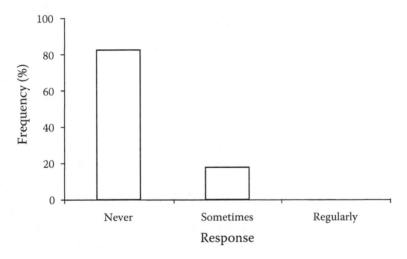

**FIGURE 6.15**   Member State use of additional aquatic macrophyte data to reduce the TER.

*Q8: To Member States and Industry: have you ever requested, evaluated (MS), or submitted (I) a modified laboratory* Lemna *study that used nonstandard environmental parameters (e.g., lower temperature)?*

No Member State has asked a notifier to submit a modified Tier 1 *Lemna* study. The few MS responses indicated that modified studies inferred modifying exposure, for example, pulsed dosing or addition of sediment. Modifications to study exposure conditions (such as lowered temperature) have been conducted by some companies, which have resulted in altered toxicity to *Lemna*. It is evident that, while there is potential value in modified *Lemna* studies, there is limited experience with them in the context of risk assessment. There is no clear opinion by Member States as to their acceptability (Figure 6.16).

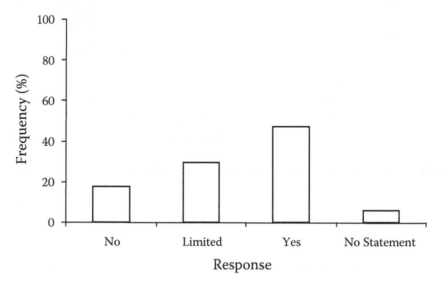

**FIGURE 6.16** Member State views on whether they would use modified *Lemna* studies in ecological risk assessment.

*Q9: Is there a need for additional or more suitable methods, species, etc.?*

The majority of MS (71%) would welcome additional guidance on methods and standardization of species (Figure 6.17). *Myriophyllum spicatum* is the macrophyte most commonly known. Differing industry views are evident, but some companies see value in higher-tier macrophyte data (where there is uncertainty at Tier 1) to inform risk assessment (Figure 6.17). Single-species tests are generally preferred. All stakeholders want reliable and appropriate test methods.

*Q10: What evaluation criteria would you apply to additional aquatic macrophyte tests?*

This question is a critical factor for all stakeholders, and a range of opinion was expressed regarding acceptability of appropriateness of vegetative growth, biomass, photosynthetic activity, and biochemical indicators as endpoints. Growth rate and biomass are the most commonly measured endpoints. There was also general agreement that in any macrophyte test there should be measurable growth in controls and a low coefficient of variation in the parameters assessed. For some, there is a question about the reliability of data from non-exponentially growing plants to assess potential for recovery.

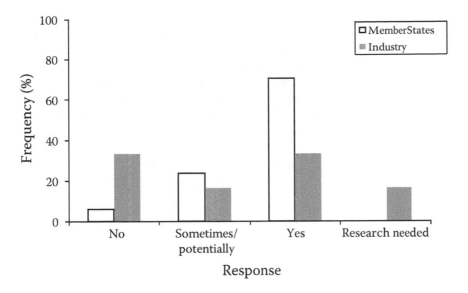

FIGURE 6.17   Member State and industry responses to Question 9.

### 6.5.3   RESPONSE FROM THE US ENVIRONMENTAL PROTECTION AGENCY: RISK ASSESSMENT FOR AQUATIC PLANTS

The risk assessment for aquatic plants in the US is different from that in Europe. In the US, 5 aquatic plant species have to be tested at the first tier of testing, four of which are algae species, plus *Lemna gibba*. For this purpose, the Office of Prevention, Pesticide and Toxic Substances (OPPTS) published draft guidelines that include two guidelines on *Lemna* (OPPTS 850.4400: Tier I + II laboratory tests, and OPPTS 850.4450: Tier III). In Tier III, field tests have to be conducted to evaluate adverse effects on sensitive native plants in ecosystems. However, few Tier III aquatic plant test have been submitted to the USEPA. In 2001, the Pest Management Regulatory Agency of Canada (PMRA) and the USEPA proposed a 4-tier scheme including additional tests with aquatic rooted macrophytes; it was presented to the Scientific Advisory Panel (Davy et al. 2001), but this scheme was never implemented.

### 6.5.4   CONCLUSIONS

From a European Member State perspective, there is a desire for improved guidance concerning the use of additional aquatic macrophyte data in risk assessment. The view of the majority of Member States and researchers is that the current data requirements are considered to be generally insufficient to assess the risk of a plant protection product to non-target aquatic macrophytes according to Directive 91/414 EEC (EU 1997). However, because only a few regulatory agencies have requested or reviewed additional aquatic macrophyte data, experience among Member States in using such data in decision-making is limited. The general view of industry is that the laboratory study using *Lemna* is an adequate test to answer aquatic macrophyte toxicity questions at Tier 1. There are different views among all stakeholders on additional data requirements, which types of tests and test species to use, and the reliability of endpoints in terms of variability in data inherent in different measurements, and how such data should be interpreted in risk assessment. Despite this uncertainty, the majority consider that effects data from non-standard macrophyte tests with additional species would improve aquatic risk assessments and could be used to lower the assessment factor. In this context, there was considerable support for a workshop to debate the issue.

Note: EU Member State and research institute data referred to above was abstracted from a UK Defra PSD (now Chemicals Regulation Directorate [CRD]) funded project, PS 2329, subject to Crown Copyright and reprinted with permission.

# References

Abernethy VJ, Sabbatini MR and Murphy KJ. (1996). Response of *Elodea canadensis* Michx. and *Myriophyllum spicatum* L. to shade, cutting and competition in experimental culture. *Hydrobiologia* 340:219–224.

Arts GHP, Belgers JDM, Hoekzema CH and Thissen JTNM. (2008). Sensitivity of submersed freshwater macrophytes and endpoints in laboratory toxicity tests. *Environ Pollut* 153:199–206.

Arts GHP, van de Hoek Tj.H, Schot JA, Sinkeldam JA and Verdonschot PFM. (2001). Biotic responses to eutrophication and recovery in outdoor experimental ditches. *Verh Int Ver Limnol* 27:1–5.

[ASTM] American Society for Testing and Materials. (2007). Standard guide for conducting static, axenic, 14-day phytotoxicity tests in test tubes with the submersed aquatic macrophyte, *Myriophyllum sibiricum* Komarov. ASTM E 1913 – 04, In: ASTM Book of Standards Volume 11.06, American Society for Testing and Materials, West Conshohocken, PA.

Barko JW and Smart RM. (1981a). Sediment-based nutrition of submersed macrophytes. *Aquat Bot* 10:339–352.

Barko JW and Smart RM. (1981b). Comparative influences of light and temperature on the growth and metabolism of selected submersed freshwater macrophytes. *Ecol Monogr* 51:219–235.

Berghahn R, Mohr S, Feibicke M, Meinecke S and Sperfeld E. (2006). Floating leaves of *Potamogeton natans*: A new method to evaluate the development of macrophytes in pond mesocosms. *Environ Sci Pollut Res* 14(3):190–193.

Biggs J, Williams P, Whitfield M, Nicolet P and Weatherby A. (2005). 15 years of pond assessment in Britain: results and lessons learned from the work of Pond Conservation. *Aquat Conserv: Marine Freshw Ecosyst* 15:693–714.

Brock TCM, Brown CD, Capri E, Gottesbüren BFF, Heimbach F, Lythgo CM, Schulz R and Streloke M (in press). Linking Aquatic Exposure and Effects in the Risk Assessment of Plant Protection Products (ELINK). SETAC and CRC Press.

Brock TCM, Lahr J and Van den Brink PJ. (2000). Ecological risks of pesticides in freshwater ecosystems, Part I: Herbicides. Alterra-Rapport 088, Wageningen, Alterra, Green World Research, The Netherlands.

Brock, TCM, Arts, GHP, Maltby L and PJ Van den Brink (2006). Aquatic Risks of Pesticides, Ecological Protection Goals and Common Aims in European Union Legislation. *Integr Environ Assess Manag* 2(4):20–46.

Brown CD, Turner N, Hollis J, Bellamy P, Biggs J, Williams P, Arnold D, Pepper T and Maund S. (2006). Morphological and physico-chemical properties of British aquatic habitats potentially exposed to pesticides. *Agric Ecosyst Environ* 113(1–4):307–319.

Campbell PJ, Arnold DJS, Brock TCM, Grandy NJ, Heger W, Heimbach F, Maund SJ and Streloke M. (1999). Guidance document on higher-tier aquatic risk assessment for pesticides (HARAP). Brussels (BE): SETAC-Europe. 179 p.

Caquet T, Lagadic L and Sheffield SR. (2000). Mesocosms in ecotoxicology (I): outdoor aquatic system. *Rev Environ Contam Toxicol* 165:1–38.

Cedergreen N, Streibig JC and Spliid NH. (2004a). Sensitivity of aquatic plants to the herbicide metsulfuron-methyl. *Ecotoxicol Environ Safety* 57:153–161.

Cedergreen N, Streibig JC and Spliid NH. (2004b). species specific sensitivity of aquatic macrophytes towards herbicides. *Ecotoxicol Environ Safety* 58:314–323.

Cedergreen N and Streibig JC. (2005). The toxicity of herbicides to non-target aquatic plants and algae: Assessment of predictive factors and hazard. *Pest Manag Sci* 61:1152–1160.

Coors A, Kuckelkorn J, Hammer-Wirtz M and Strauss T. (2006). Application of in-situ bioassays with macrophytes in aquatic mesoscosm studies. *Ecotoxicology* 15:583–591.

Davies B, Biggs J, Williams P, Whitfield M, Nicolet P, Sear D, Bray S and Maund S. (2008). Comparative biodiversity of aquatic habitats in the European agricultural landscape *Agric Ecosyst Environ* 125:1–8.

Davies J, Honegger JL, Tencalla FG, Meregalli G, Brain P, Newman JR and Pitchford HF. (2003). Herbicide risk assessment for non-target aquatic plants: sulfosulfuron — a case study. *Pest Manag Sci* 59(2):231–237.

Davies J. (2001). Guidelines for assessing the effects of pesticides on the growth of *Glyceria maxima*. Protocol available from PSD, DEFRA, UK.

Davy M, Petrie R, Smrchek J, Kuchnicki T and Francois D. (2001). Proposal to update non-target plant toxicity testing under NAFTA. United States Environmental Protection Agency. http://www.epa.gov/scipoly/sap/meetings/2001/june/sap14.pdf (accessed July 2008).

De Jong FMW, Brock TCM, Foekema EM and Leeuwangh P. (2008). Guidance for summarizing and evaluating aquatic micro- and mesocosm studies. RIVM Report 601506009, Bilthoven, The Netherlands, 59 p.

De Zwart D. (2002). Observed regularities in species sensitivity distributions for aquatic species. In Posthuma L, Suter GWI, Traas TP (Eds), Species-Sensitivity Distributions in Ecotoxicology. Lewis Publishers, Boca Raton, FL. p 133–154.

[EC] European Commission. (2002). Guidance Document on Aquatic Ecotoxicology in the context of Directive 91/414/EEC. Sanco/3268/2001 rev 4 (final). European Commission Health & Consumer Protection Directorate-General, Brussels.

Elendt BP. (1990). Selenium deficiency in crustacea; an ultrastructural approach to antennal damage in *Daphnia magna* Straus. *Protoplasma* 154:25–33.

[EU] European Union. (1997). Commission proposal for a council directive establishing Annex VI to Directive 91/414/EEC concerning the placing of plant protection products on the market. Off J Eur Comm C 240:1–23.

Feiler U, Kirchesch I and Heininger P. (2004). A new plant-based bioassay for aquatic sediments. *J Soils Sediments* 4:261–266.

Friesen LJ Shane, Van Eerd Laura L, Hall Christopher. (2003). Herbicides, Plant Hormone Mimics — Auxins. Encyclopedia of Agrochemicals, John Wiley & Sons, New York.

Giddings JM, Brock TCM, Heger W, Heimbach F, Maund SJ, Norman S, Ratte H-T, Schäfers C and Streloke M, editors. (2002). Community-level Aquatic System Studies–Interpretation Criteria. Pensacola (FL): SETAC. 44 p.

Hanson ML, Sanderson H and Solomon KR. (2003). Variation, replication, and power analysis of *Myriophyllum* spp. microcosm toxicity data. *Environ Toxicol Chem* 22:1318–1329.

Hanson ML, Sibley PK, Brain RA, Mabury SA and Solomon KR. (2005). Microcosm evaluation of the toxicity and risk to aquatic macrophytes from perfluorooctane sulfonic acid. *Arch Environ Contam Toxicol.* 48(3):329–337.

Hanson ML, Sibley PK, Ellis DA, Mabury SA, Muir DCG and Solomon KR. (2002). Evaluation of monochloroacetic acid (MCA) degradation and toxicity to *Lemna gibba*, *Myriophyllum spicatum*, and *Myriophyllum sibiricum* in aquatic microcosms. *Aquat Toxicol* 61:251–273.

Hanson ML, Sibley PK, Solomon KR, Mabury SA and Muir DCG. (2001). Chlorodifluoroacetic acid (CDFA) fate and toxicity to the macrophytes *Lemna gibba*, *Myriophyllum spicatum* and *Myriophyllum sibiricum* in aquatic microcosms. *Environ Toxicol Chem* 20:2758–2767.

Huber W. (1994). Atrazine in aquatic test systems: an evaluation of ecotoxicological risks. In: Hill, IR, Heimbach F, Leeuwangh P, Matthiessen P (Eds.). Freshwater Field Tests for Hazard Assessment of Chemicals. CRC Press, Boca Raton, FL.

Huebert DB and Shay JM. (1993). Considerations in the assessments of toxicity using duckweeds. *Environ Toxicol Chem* 12:481–483.

Huggins DG, Johnson ML and deNoyelles F. (1993). The ecotoxic effects of atrazine on aquatic ecosystems: An assessment of direct and indirect effects using structural equation modelling. In: Graney RL, Kennedy JH, Rodgers Jr. JH (Eds). Aquatic Mesocosm Studies in Ecological Risk Assessment. Lewis Publishers, Boca Raton, FL. p 653–692.

ISO Standard No. 8692 (2004). Water quality–freshwater algal growth inhibition test with unicellular green algae.

Jones PJ, Curtis C, Moss B, Whitehead P, Bennion H and Patrick S. (1998). *Trial classification of lake water quality in England and Wales: a proposed approach*. R&D Technical Report E53S, Environment Agency Publications, Bristol, UK, 80 pp.

Knauer K, Mohr S and Feiler U. (2008). Comparing growth development of *Myriophyllum* spp. in laboratory and field experiments for ecotoxicological testing. *Environ Sci Pollut Res* 15:322–331.

Knauer K, Vervliet-Schneebaum M, Dark RJ and Maund SJ, (2006). Methods for assessing the toxicity of herbicides to submersed aquatic plants. *Pest Manage Sci* 62:715–722.

Kubitza J and Dohmen GP. (2002). Effect of metazachlor on submersed and emergent aquatic plants. Poster, SETAC-Europe meeting, Vienna 2002.

Kubitza J and Dohmen GP. (2008). Development of a test method for the aquatic macrophyte *Myriophyllum aquaticum*. SETAC Europe 18th Annual Meeting, Warsaw, 2008.

Küster A and Altenburger R. (2007) Development and validation of a new fluorescence-based bioassay for aquatic macrophyte species. *Chemosphere* 67:194–201.

Laurie EMO. (1948). The dissolved oxygen of an upland pond and its inflowing stream, at Ystumtuen, North Cardiganshire, Wales. *J Ecol* 30: 357–382.

Madsen JD, Hartleb CF and Boylen CW. (1991). Photosynthetic characteristics of *Myriophyllum spicatum* and six submersed aquatic macrophyte species native to Lake George, New York. *Freshw Biol* 26:233–240.

Maletzki D, Beulshausen T, Feibicke M and Kussatz C. (2008). Sediment free testing on aquatic macrophyte *Myriophyllum spicatum*. Poster Presentation at the 18th Annual Meeting of SETAC Europe May 2008 in Warsaw, Poland.

Maltby L, Blake N, Brock TCM and Van Den Brink PJ. (2005). Insecticide species sensitivity distributions: importance of test species selection and relevance to aquatic ecosystems. *Environ Toxicol Chem* 24:379–388.

McGregor EB, Solomon KR and Hanson ML. (2007). Monensin is not toxic to aquatic macrophytes at environmentally relevant concentrations. *Arch Environ Contam Toxicol* 53(4):541–551.

McGregor EB, Solomon KR and Hanson ML. (2008). Effects of planting system design on the toxicological sensitivity of *Myriophyllum spicatum* and *Elodea canadensis* to atrazine. *Chemosphere* 73(3):249–260.

Mieczan T. (2007). Comparative study of periphytic ciliate communities and succession on natural and artificial substrata in two shallow lakes (Eastern Poland). *Ann Limnol-Int J Limnol* 43:179–186

Mohr S, Berghahn R, Feibicke M, Meinecke S, Ottenströer T, Schmiedling I and Schmidt R. (2007). Effects of the herbicide metazachlor on macrophytes and ecosystem function in freshwater pond and stream mesocosms. *Aquat Toxicol* 82(2007):73–84.

Moody, ML and Les DH. (2002). Evidence of hybridity in invasive watermilfoil (*Myriophyllum*) populations. *PNAS* 99(23):14867–71.

Nijboer RC. (2004). Literatuurstudie naar hydrologische maatregelen en de effecten op sloot- en beekecosystemen. Alterra report 1066 (in Dutch).

[OECD] Organisation for Economic Co-operation and Development. (2002). Freshwater alga and cyanobacteria, growth inhibition test. OECD 201 Guidelines for the Testing of Chemicals.

[OECD] Organisation for Economic Co-operation and Development. (2006a). Guidance Document on Simulated Freshwater Lentic Field Tests (Outdoor Microcosms and Mesocosms) (http://www.oecd.org/dataoecd/30/10/32612239.pdf) (accessed July 2008).

[OECD] Organisation for Economic Co-operation and Development. (2006b). Guidelines for the Testing of Chemicals No. 201 – Freshwater Alga and Cyanobacteria, Growth Inhibition Test. Adopted: 23 March 2006. OECD, Paris.

[OECD] Organisation for Economic Co-operation and Development. (2006c). OECD Guidelines for the Testing of Chemicals No. 221: *Lemna* sp. Growth Inhibition Test. Adopted: 23 March 2006. OECD, Paris.

[OECD] Organisation for Economic Co-operation and Development. (2004a). OECD Guidelines for the Testing of Chemicals No. 218: Sediment-Water Chironomid Toxicity Using Spiked Sediment. Adopted: 13 April 2004. OECD, Paris.

[OECD] Organisation for Economic Co-operation and Development. (2004b). OECD Guidelines for the Testing of Chemicals No. 219: Sediment-Water Chironomid Toxicity Using Spiked Water. Adopted: 13 April 2004. OECD, Paris.

O'Hare MT, Stillman RA, Mcdonnell J and Wood LR. (2007). Effects of mute swan grazing on a keystone macrophyte. *Freshw Biol* 52:2463–2475.

Poovey AG and Getsinger KD. (2005). Development of a standard growth protocol for the submersed macrophyte *Myriophyllum spicatum* to test non-target effects of chemicals in aquatic systems. USA SETAC – Baltimore, November 2005.

Preston CD, Pearman DA and Dines TD. (2002). New Atlas of the British and Irish Flora. Oxford: Oxford University Press.

Rentz N, Solomon KR and Hanson ML. (2009). Investigating the influence of relative growth rate, population density and stand establishment on toxicity in macrophytes: A case study with diuron and *Elodea canadensis*. *Environ Toxicol Chem* (in press).

Roshon RD, Stephenson GR and Horton RF. (1996). Comparison of five media for the axenic culture of *Myriophyllum sibiricum* Komarov. *Hydrobiologia* 340:17–22.

Sanderson H, Laird B, Brain R, Wilson CJ and Solomon KR. (2008). Yearly seasonal ecological development and biometry (null hypothesis and bioequivalence testing) of 15 untreated 12 m³ aquatic mesocosms. *Ecotoxicology* (accepted).

Shaw JL and Kennedy JH. (1996). The use of aquatic field mesocosm studies in risk assessment. *Environ Toxicol Chem* 15:605–607.

Skogerboe JG, Getsinger KD, and Glomski LM. (2006). Efficacy of diquat on submersed plants treated under simulated flowing water. *J Aquat Plant Manag* 44:122–125.

Smart RM and Barko JW. (1985). Laboratory culture of submersed freshwater macrophytes on natural sediments. *Aquat Bot* 21:251–263.

Solomon KR. (1996). Overview of recent developments in ecotoxicological risk assessment. *Risk Analysis* 16:627–633.

Solomon KR, Brock TCM, De Zwart D, Dyer S, Posthuma L and Richardset S. (2008). Extrapolation Practice for Ecotoxicological Effect Characterization of Chemicals. SETAC Press and CRC Press, Boca Raton, FL. p 257–267.

Spencer DF and Ksander GG. (2005). Root size and depth distribution for three species of submersed aquatic plants grown alone or in mixtures: Evidence for nutrient competition. *J Freshw Ecol* 20:109–116.

Technical Report E53. Bristol, Environment Agency. Can be downloaded at: http://publications. environment-agency.gov.uk/epages/eapublications.storefront.

[USEPA] US Environmental Protection Agency. (1971). Algal Assay Procedure Bottle Test. US Environmental Protection Agency, Corvallis, Oregon.

[USEPA] US Environmental Protection Agency. (1996). Ecological Effects Test Guidelines No. EPA 712–C–96–156: OPPTS 850.4400, Aquatic Plant Toxicity Test Using *Lemna* spp., Tiers I and II.

[UBA] Umweltbundesamt. (2007). Field, fate and effect studies on the biocide N-tert-butyl-N'-cyclopropyl-6-methylthio-1,3,5-triazine-2,4-diamine Irgarol (CAS-No 28159-98-0). Umweltbundesamt, report, May 2007.

Van Den Brink PJ, Blake N, Brock TCM, and Maltby L. (2006). Predictive value of species sensitivity distributions for effects of herbicides in freshwater ecosystems. *Human and Ecological Risk Assessment* 12:645–674.

Van Liere L, Janse JH and Arts GHP. (2007). Setting nutrient values for ditches using the eutrophication model PC DITCH. *Aquat Ecol* 41(3):443–449.

Van Vlaardingen PLA, Traas TP, Wintersen AM, Aldenberg T. (2004). ETX 2.0. A Program to Calculate Hazardous Concentrations and Fraction Affected, Based on Normally Distributed Toxicity Data. RIVM The Netherlands.

Van Wijngaarden RPA, Brock TCM, van den Brink PJ, Gylstra R and Maund SJ. (2006). Ecological effects of spring and late summer applications of lambda-cyhalothrin on freshwater microcosms. *Arch Environ Contam Toxicol* 50(2):220–239.

Van Wijngaarden RPA, Cuppen JGM, Arts GHP, Crum SJH, van den Hoorn MW, van den Brink PJ and Brock TCM. (2004). Aquatic risk assessment of a realistic exposure to pesticides used in bulb crops: A microcosm study. *Environ Toxicol Chem* 23(6):1479–1498.

Vervliet-Scheebaum M, Knauer K, Maund SJ, Grade R and Wagner E. (2006). Evaluating the necessity of additional plant testing by comparing the sensitivities of different species. *Hydrobiologia* 570:231–236.

White J and Irvine K. (2003). The use of littoral mesohabitats and their macroinvertebrate assemblages in the ecological assessment of lakes. *Aquatic Conservation: Marine Freshw Ecosyst* 13(4):331–351.

Williams P, Whitfield M, Biggs J, Bray S, Fox G, Nicolet P and Sear D. (2004). Comparative biodiversity of rivers, streams, ditches and ponds in an agricultural landscape in Southern England. *Biol Conserv* 115:329–341.

# Appendix I: AMRAP Case Studies

## A1.1 CASE STUDY A: AMRAP-AUXIN

### A1.1.1 GENERAL PROPERTIES

| | **Identity** |
|---|---|
| Chemical class: | Aryloxyalkanoic acid |
| Mode of action: | Synthetic auxin; interacts with indoleacetic acid, ethylene and abscisic acid pathways leading to growth deformities, growth inhibition, and ultimately senescence. |
| Uptake and translocation properties: | Uptake via foliage and roots; translocation occurs acropetally and basipetally in the phloem. |

| | **Selectivity** |
|---|---|
| Crop: | Spring cereals |
| Target weeds: | Broad-leaved weeds |
| Mechanism of selectivity: | Selectivity between dicots and grasses is believed to result from morphological differences and differential rates of translocation and metabolism. Auxinic herbicides cause abnormal tissue proliferation, which destroys the phloem of dicot species. In grass species, the phloem bundles are scattered within stem ground tissue and are surrounded by protective sclerenchyma tissue. Also, the pericycle of vascular bundles in grasses is not sensitive to auxinic herbicides (Friesen et al. 2003). |

| | **Use Pattern** |
|---|---|
| Timing of application: | BBCH 15 to 31 |
| Application rate: | Single application of 1200 g ai/ha |

| **Physico-Chemical and Environmental Fate Properties** | |
|---|---|
| Molecular weight: | 220 g/mol |
| Water solubility: | 24 mg/L at 20 °C, pH 7 |
| Log $P_{OW}$: | –0.75 at 20 °C, pH 7 |
| Soil $t\frac{1}{2}$: | Mean from laboratory studies 13.5 days |
| Water/sediment $t\frac{1}{2}$: | Whole system 31 days |
| $K_{OC}$: | 60 (mean) |
| 1/n: | 0.9 (default) |

### A1.1.2 Predicted Environmental Concentrations

Predicted environmental concentrations of the active substance in surface water and sediment were estimated using the FOCUS Surface Water model based on the input parameters shown above. The maximum resulting PEC values are presented in Table A1.1.

### A1.1.3 Tier 1 Toxicity Data for Algae and Aquatic Plants

In accordance with Directive 91/414/EEC, toxicity tests were conducted with algal species and the higher aquatic plant, *Lemna gibba*. The Guidance Document on Aquatic Ecotoxicology (EC 2002) states that where there is evidence from efficacy data on terrestrial plants that the data for *Lemna* are not representative for other aquatic plant species (e.g., auxin simulators, which can be more toxic to submerged plants than to *Lemna*), additional data with other aquatic plant species may be required on a case-by-case basis. Therefore, in addition to standard Tier 1 tests with algae and *Lemna* species, the effects of this auxinic herbicide were also evaluated in the submerged, rooted macrophyte, *Myriophyllum spicatum* (Study 1). Endpoints from all studies are presented in Table A1.2.

### A1.1.4 Toxicity Exposure Ratios

The risk to algae and aquatic plants was evaluated by calculation of toxicity exposure ratios (TERs) based on the maximum initial PEC generated from FOCUS SW Step 3 and the lowest EC50 and NOEC endpoints for each species (Table A1.3).

### A1.1.5 Higher-Tier Data: Laboratory Studies with Additional Species (Studies 2 and 3)

In addition to the *Myriophyllum spicatum* study reported overleaf (Study 1), two further laboratory studies were conducted with submerged macrophyte species. In

---

### TABLE A1.1
### FOCUS SW Step 3 PEC values at 1 m from edge of treated area

| Scenario | Waterbody | Surface Water PEC (µg ai/L) Maximum Initial | 7-d TWA | Sediment PEC (µg ai/kg) Maximum Initial | 7-d TWA |
|---|---|---|---|---|---|
| D1 | Ditch | 8.60 | 3.41 | 4.37 | 3.96 |
| D1 | Stream | 6.62 | 1.07 | 2.11 | 2.10 |
| D3 | Ditch | 7.61 | 1.06 | 1.27 | 0.679 |
| D4 | Pond | 0.270 | 0.254 | 0.336 | 0.336 |
| D4 | Stream | 5.92 | 0.055 | 0.206 | 0.056 |
| D5 | Pond | 0.273 | 0.254 | 0.336 | 0.336 |
| D5 | Stream | 5.88 | 0.035 | 0.137 | 0.037 |
| R4 | Stream | 5.01 | 0.132 | 0.382 | 0.094 |

## TABLE A1.2
### Tier 1 Algal and aquatic plant endpoints used for risk assessment

| Species | Test Substance | Test Duration | Exposure Regime | EC50* (mg ai/L) | NOEC (mg ai/L) |
|---|---|---|---|---|---|
| Pseudokirchneriella subcapitata (green alga) | Active ingredient | 72 hours | Static | 41.0 | 10.0 |
| Pseudokirchneriella subcapitata | Formulation | 72 hours | Static | 24.0 | 5.0 |
| Lemna gibba (gibbous duckweed) | Active ingredient | 7 days | Static renewal (renewal on days 3 and 5) | 0.580 | 0.290 |
| Myriophyllum spicatum (Eurasian watermilfoil) | Formulation | 21 days | Static (concentrations remained within 10% of nominal over the test duration) | 0.0125 | 0.0045 |

* Most sensitive endpoint based on biomass or growth rate

## TABLE A1.3
### Algae and *Lemna* TER values for FOCUS Step 3 scenarios

| Species | Endpoint | Endpoint Value (mg ai/L) | Maximum Initial PEC (µg ai/L) | TER (2 Sig. Figures) |
|---|---|---|---|---|
| Pseudokirchneriella subcapitata | 72-h EC50 | 24.0 | | 2800 |
| | 72-h NOEC | 5.0 | | 580 |
| Lemna gibba | 7-d EC50 | 0.580 | 8.6 | 67 |
| | 7-d NOEC | 0.290 | | 34 |
| Myriophyllum spicatum | 21-d EC50 | 0.0125 | | 1.5 |
| | 21-d NOEC | 0.0045 | | 0.52 |

the first of these studies (Study 2), *M. aquaticum* was evaluated in a test system containing a rooting substrate. In a further study (Study 3), 9 submerged species, including *M. spicatum*, were tested in the absence of sediment. However, endpoints were generated for only 8 species due to lack of growth in *M. spicatum* control cultures. Endpoints from the standard Tier 1 *Lemna* test and Studies 1, 2, and 3 are summarized in Table A1.4.

Species sensitivity distributions (SSDs) based on EC50 endpoints were prepared for each assessment parameter using the program ETX 2.0 (Van Vlaardingen et al. 2004). All available endpoints were included, although EC50 values lying outside the exposure range of the test (i.e., greater-than values) were omitted. The resulting median, upper, and lower HC5 values (i.e., hazard concentration affecting 5% of species) are presented in Table A1.4.

**TABLE A1.4**
**Summary of endpoints from additional species tests**

| | | | | EC50 (µg ai/L) | | | | |
|---|---|---|---|---|---|---|---|---|
| Study Number | Species | Species Classification | Test Duration (Days) | Shoot (Dry Weight) | Root (Dry Weight) | Shoot Length | Root Length | Number of Roots |
| Tier 1 study | *Lemna gibba* (gibbous duckweed) | Floating, non-rooted monocot | 7 | 580 | n.d. | n.d. | n.d. | n.d. |
| 1 | *Myriophyllum spicatum* (Eurasian water milfoil) | Submerged rooted dicot | 21 | 64.2 | n.d. | 12.5 | n.d. | n.d. |
| 2 | *Myriophyllum aquaticum* (Northern water milfoil) | Submerged rooted dicot | 14 | 1470 | n.d. | n.d. | 50 | 158 |
| 3 | *Lemna trisulca* (ivy-leaved duckweed) | Submerged, non-rooted monocot | 28 | >3000 | n.d. | n.d | n.d. | n.d. |
| 3 | *Ceratophyllum demersum* (coontail) | Submerged non-rooted dicot | 28 | >3000 | >3000 | >3000 | n.d. | n.d. |
| 3 | *Elodea nuttallii* (Nutall's waterweed) | Submerged rooted monocot | 28 | 2243 | 898 | 785 | 574 | 982 |
| 3 | *Ranunculus aquatilis* (common water crowfoot) | Submerged rooted dicot | 28 | >3000 | n.d. | 342 | 108 | n.d. |

| | | | | | | | | |
|---|---|---|---|---|---|---|---|---|
| 3 | *Ranunculus circinatus* (fan-leaved watercrowfoot) | Submerged rooted dicot | 28 | 2731 | 111 | 1120 | 100 | 112 |
| 3 | *Ranunculus peltatus* (pond water crowfoot) | Submerged rooted dicot | 28 | n.d. | 245 | 140 | 263 | 271 |
| 3 | *Potamogeton lucens* (shining pondweed) | Submerged rooted monocot | 28 | >3000 | n.d. | 1063 | 181 | 299 |
| 3 | *Potamogeton crispus* (curly pondweed) | Submerged rooted moncot | 28 | >3000 | 347 | 1988 | 290 | 326 |
| **Median HC5 (µg ai/L)** | | | | **51.6** | **62.2** | **18.6** | **42.0** | **75.5** |
| **Lower–upper limit HC5 (µg ai/L)** | | | | **1.25–226** | **3.5–159** | **1.02–76.3** | **10.9–81.1** | **17.6–145** |
| *n* | | | | **5** | **4** | **7** | **7** | **6** |

*n.d. not determined*

## TABLE A1.5
**Effects of an auxin herbicide on *Myriophyllum* sp. and *Potamogeton* sp. in an outdoor microcosm study**

| | | 30 Days After Treatment | | 60 Days After Treatment | |
|---|---|---|---|---|---|
| Species | Exposure Concentration (mg ai/L) | Mean Total Plant Fresh Weight (g Per Plant) | Plants Showing Symptoms (%) | Mean Total Plant Fresh Weight (g Per Plant) | Plants Showing Symptoms (%) |
| *Potamogeton sp.* | 0 | 5.6 | 0 | 8.5 | 34 |
| | 0.01 | 6.9 | 0 | 13.9* | 23 |
| | 0.1 | 4.2 | 62* | 5.4* | 89* |
| *Myriophyllum sp.* | 0 | 14.5 | 0 | 42.6 | 9 |
| | 0.01 | 16.2 | 0 | 53.9 | 0 |
| | 0.1 | 11.0 | 70* | 28.6* | 90* |

* indicates significant difference from corresponding control

## A1.1.6 STUDY 4: OUTDOOR MICROCOSM STUDY

A microcosm study (Study 4) was conducted to evaluate the effects of the test substance on the growth of submerged *Myriophyllum* and *Potamogeton* species. Outdoor enclosures were filled with a layer of natural sediment overlaid with natural pond water. Young plants with roots were collected from natural ponds and transplanted into plastic pots containing natural sediment, which were placed on the surface on the sediment in each enclosure. The study incorporated 3 replicate enclosures per treatment, each containing 12 individually potted plants per species. The test substance was applied by mixing into the water column to give nominal concentrations of 0.01 and 0.1 mg ai/L. Assessments of plant fresh weight and the number of plants exhibiting symptoms of toxicity were made 30 and 60 days after treatment (Table A1.5). The NOEC for both species was 0.01 mg ai/L.

## A1.2 CASE STUDY B: AMRAP-PHENYLUREA

### A1.2.1 GENERAL PROPERTIES

**Identity**

| | |
|---|---|
| Chemical class: | Substituted phenylurea herbicide |
| Mode of action: | Photosynthesis (PSII) inhibitor |
| Uptake and translocation properties: | Herbicide is systemic and absorbed principally by the roots but also by foliage, with translocation occurring acropetally in the xylem. |

**Selectivity**

| | |
|---|---|
| Crop: | Root vegetables (carrots) |
| Target weeds: | Broad-leaved weeds and annual grasses |

*(Continued)*

**Use Pattern**

| | |
|---|---|
| Timing of application: | Post-emergence BBCH 12 |
| Application rate: | Single application of 1000 g ai/ha |

**Physico-Chemical and Environmental Fate Properties**

| | |
|---|---|
| Molecular weight: | 249 g/mol |
| Water solubility: | 68.3 mg/L at 20 °C, pH 7 |
| Log $P_{ow}$: | 3.0 at 20 °C, pH 7 |
| Soil t½: | 38–92 days laboratory; 15–82 days field |
|   Worst-case t½ of 92 days used in FOCUS SW modeling | |
| Water/sediment t½: | Whole system: 46 days in silt loam; 230 days in sandy loam |
| | Water: 48 days in silt loam; 220 days in sandy loam |
| | Whole system t½ of 46 days used in FOCUS SW modeling |
| $K_{OC}$: | 450 |
| 1/n: | 0.79–0.91 (0.9 used in FOCUS SW modeling) |
| Hydrolysis DT50 at pH 5, 7, 9: | >1000 days |
| Photolysis DT50: | >260 days |

## TABLE A1.6
## FOCUS SW Step 3 PEC values at 1 m from edge of treated area

| Scenario | Waterbody | Surface Water PEC (µg ai/L) | | Sediment PEC (µg ai/kg) | |
|---|---|---|---|---|---|
| | | Maximum Initial | 7-d TWA | Maximum Initial | 7-d TWA |
| D3 | Ditch | 6.327 | 0.883 | 2.523 | 1.469 |
| D6 | Ditch | 6.640 | 2.061 | 4.504 | 4.357 |
| R1 | Pond | 1.703 | 1.644 | 6.327 | 6.314 |
| R1 | Stream | 15.58 | 2.419 | 9.128 | 6.238 |
| R2 | Stream | 5.605 | 0.664 | 2.226 | 1.819 |
| R2 | Stream | 5.618 | 1.115 | 13.53 | 11.21 |
| R3 | Stream | 17.58 | 1.277 | 5.084 | 2.367 |
| R4 | Stream | 24.70 | 2.560 | 10.50 | 4.897 |

### A1.2.2 PREDICTED ENVIRONMENTAL CONCENTRATIONS

Predicted environmental concentrations of the active substance in surface water and sediment were estimated using the FOCUS Surface Water model based on the input parameters shown above. The resulting PEC values are presented in Table A1.6.

### A1.2.3 TIER 1 TOXICITY DATA FOR ALGAE AND AQUATIC PLANTS

Standard Tier 1 algal and aquatic plant studies were conducted in accordance with internationally recognized guidelines, using species recognized as appropriate test species at the time, and in accordance with GLP. Endpoints from these studies are presented in Table A1.7.

## TABLE A1.7
## Tier 1 Algal and aquatic plant endpoints used for risk assessment

| Species | Test Substance | Test Duration | Exposure Regime | EC50 (µg ai/L) | NOEC (µg ai/L) |
|---|---|---|---|---|---|
| *Pseudokirchneriella subcapitata* (green alga) | Active substance | 72 hours | Static | 16.0 | 5.6 |
| *Chlorella vulgaris* (green alga) | Active substance | 7 days | Static | 7 | Not determined |
| *Lemna minor* (duckweed) | Active substance | 7 days | Static | 7 | Not determined |

### A1.2.4    PRELIMINARY TOXICITY EXPOSURE RATIOS

The risk to algae and aquatic plants was evaluated by calculation of toxicity exposure ratios (TERs) based on the maximum initial and/or 7-day TWA surface water PEC value generated from FOCUS SW Step 3 (Table A1.8).

### A1.2.5    HIGHER-TIER DATA

### A1.2.5.1    Study 1: Laboratory Study with *Myriophyllum spicatum* and *Potamogeton perfoliatus*

A higher-tier laboratory study was conducted with 2 submerged macrophyte species, *Myriophyllum spicatum* and *Potamogeton perfoliatus*. Glass aquaria containing a layer of natural sediment overlaid with natural pond water were planted with 10 shoots of each species. After 7 weeks, microcosms were treated with a single application of the test substance and maintained for a further 5 weeks. Assessments of shoot biomass were made at test initiation and test termination, and photosynthesis, measured as oxygen evolution, was recorded twice weekly.

After 5 weeks, *P. perfoliatus* biomass significantly decreased with increasing herbicide concentrations of 50 µg ai/L and above, with a non-statistically significant increase at 5 µg ai/L. Significant reductions in *M. spicatum* biomass were seen at 500 and 1000 µg ai/L. Dose–response curves were fitted to photosynthetic and biomass data to allow calculation of EC50 values (Table A1.9). Comparison of biomass EC50 values indicates that *P. perfoliatus* is about 6× more sensitive than *M. spicatum*.

### A1.2.5.2    Study 2: Indoor Mesocosm and *Elodea nuttallii* Bioassay

An indoor study was conducted on aquatic plants and algae in glass aquaria (600 L) containing a layer of natural sediment overlaid with 50 cm water. Plankton, macroinvertebrates, and *Elodea nuttallii* were added and acclimatized for 3 months prior to treatment with the test substance. Microcosms were treated twice weekly for 4 weeks followed by a 7-week non-treatment phase. Assessments of periphyton and phytoplankton abundance, species composition, and chlorophyll-a concentrations were made on alternate weeks. *Elodea* shoots were harvested after 11 weeks for assessment of fresh and dry weight.

## TABLE A1.8
## Algae and *Lemna* TER values based on maximum FOCUS Step 3 PEC values

| Species | Endpoint | Endpoint Value (µg ai/L) | Maximum Initial PEC (µg ai/L) | TER | 7-Day TWA PEC (µg ai/L) | TER |
|---|---|---|---|---|---|---|
| *Pseudokirchneriella* | 72-h EC50 | 16 | | 0.65 | n.a. | n.d. |
| *subcapitata* | 72-h NOEC | 5.6 | | 0.23 | n.a. | n.d. |
| *Chlorella vulgaris* | 7-d EC50 | 7 | 24.7 | 0.28 | 2.56 | 2.7 |
| | 7-d NOEC | n.d. | | n.d. | n.a. | n.d. |
| *Lemna minor* | 7-d EC50 | 7 | | 0.28 | 2.56 | 2.7 |
| | 7-d NOEC | n.d. | | n.d. | n.a. | n.d. |

*n.d. not determined; n.a. not applicable*

## TABLE A1.9
## Effects of the test substance on photosynthesis and biomass in *Myriophyllum spicatum* and *Potamogeton perfoliatus* under laboratory conditions

| Species | Photosynthesis ($O_2$ Evolution) | | | Final Biomass | | |
|---|---|---|---|---|---|---|
| | $r^2$ | EC50 (µg ai/L) | NOEC (µg ai/L) | $r^2$ | EC50 (µg ai/L) | NOEC (µg ai/L) |
| *P. perfoliatus* | 1.0 | 45 | 5.0 | 0.95 | 25 | 5.0 |
| *M. spicatum* | 0.94 | 80 | 5.0 | 0.93 | 137 | 100.0 |

In addition, an in-situ *Elodea* bioassay was performed whereby shoots (4 g) were allowed to root in a plastic pot containing sediment. Pots were then enclosed in a cage, within each mesocosm and shoots were harvested after 3 weeks for determination of shoot dry weight. A summary of endpoints based on *Elodea* biomass in the bioassay and the mesocosm is provided in Table A1.10.

An increase in *Chlamydomonas* abundance occurred at 50 and 150 µg ai/L. For *Cocconeis* sp. and *Phormidium foveolarum,* there was a decrease in abundance at the same concentrations. The most affected species was *Chroomonas* with a significant decrease in abundance at all concentrations but with evidence of recovery after the 7-week post-treatment period at concentrations of 15 µg ai/L and below. Analysis of periphyton on glass slides showed a dominance of *Cocconeis* sp. and *Chlamydomonas.* During the post treatment phase, *Chlamydomonas* became dominant in the 2 highest treatment rates, while *Cocconeis* sp. continued to decline in the same treatments.

### A1.2.5.3 Study 3: Outdoor Mesocosm Study

An outdoor study was conducted to evaluate the effects of the test substance in large experimental drainage ditches containing established populations of macrophytes,

**TABLE A1.10**

**Summary of effects on *Elodea nuttallii* biomass**

| Nominal Concentration (µg ai/L) | *Elodea* Bioassay Biomass Change as % of Control During the First 21 Days | *Elodea* Biomass in Mesocosm Standing Stock at Week 11 as % of Control |
|---|---|---|
| 0.5 | 93 | 130 |
| 5 | 80* | 123 |
| 15 | 66* | 99 |
| 50 | 56* | 47* |
| 150 | 42* | 5* |
| EC50 (µg ai/L) | 75 | 15 |
| NOEC (µg ai/L) | 0.5 | n.d. |

*statistically different from control*

phytoplankton, zooplankton, and macroinvertebrates. Ditches were macrophyte-dominated and were treated once every 4 weeks to give a total of 3 applications. The herbicide was applied by spray boom on to the static water surface. After 7 days, the treated water was flushed with reservoir water for 3 weeks with a residence time of 8 days. This process was repeated after each application. After the final application and subsequent 7-day static period, the flow was maintained until the end of the experiment, that is, 14 weeks after the final application. Herbicide concentrations were measured at frequent intervals between each application and at intervals after the final application until day 100 after the first application.

Macrophyte species composition and abundance were monitored at designated intervals. The percentage cover of each macrophyte species was estimated on a monthly basis from April until October. Above-ground biomass was estimated in April, June, August, and October. Assessments of periphyton and phytoplankton abundance and species composition, and chlorophyll-a concentration, were made at weekly intervals until the fourth week after the final application. Thereafter, assessments were made every other week.

The calculated DT50 was in the range of 7 to 12 days. Of the 12 macrophyte species present, the dominant species were *Sagittaria sagittifolia*, *Myriophyllum spicatum*, and *Elodea nuttallii*. *Ranunculus*, *Potamogeton*, and *Polygonum* species were also abundant in some mesocosms. *S. sagittifolia* and *M. spicatum* increased in abundance during the first 2 treatment periods, after which time *S. sagittifolia* showed signs of senescence in all mesocosms. Both *M. spicatum* and *E. nuttallii* dominated until the end of the season. No relationship between total number of macrophyte species and herbicide treatment was evident, nor was there a significant difference in mean cover of macrophytes in any treatment compared with controls (Table A1.11). There was a non-significant reduction in biomass at 50 µg ai/L after the second application.

During the course of the study, 52 taxa of phytoplankton were identified. The dominant groups were Diatomeae (represented predominantly by *Synedron*, *Achnanthes*, and *Meridion* species), Crytophyceae (represented predominantly by *Chryptomonas*

## TABLE A1.11
## Effect of the test substance on total macrophyte biomass over the test duration

| Nominal Concentration (µg ai/L) | Macrophyte Biomass (Mean of 2 Replicates; g dw/m$^2$) | | | |
|---|---|---|---|---|
| | Pre-Treatment | 2nd Week After 2nd Application | 4th Week After 3rd Application | 14th Week After 3rd Application |
| 0 | 30.6 (13.6–47.5) | 61.9 (45.0–78.8) | 152.7 (129.6–175.8) | 149.8 (97.1–202.5) |
| 0.5 | 40.4 (35.9–44.8) | 76.2 (66.0–86.4) | 146.3 (128.7–163.8) | 137.0 (122.3–151.7) |
| 5 | 15.9 (12.5–19.3) | 77.5 (63.0–92.0) | 165.1 (139.7–190.4) | 143.5 (101.3–185.6) |
| 15 | 16.8 (15.8–17.7) | 75.3 (56.7–93.8) | 139.0 (111.2–166.7) | 141.2 (92.1–190.3) |
| 50 | 20.0 (15.9–24.0) | 49.6 (35.2–64.0) | 110.1 (104.5–115.7) | 114.6 (97.8–131.4) |

species), and Chlorophyceae (represented predominantly by *Chlamydomonas* species). Multivariate analyses did not reveal significant, consistent changes in the species composition of mesocosms treated with the test substance relative to untreated mesocosms. Analysis of abundance indicated that only *Chlamydomonas* species showed consistent and significant reductions following exposure to 15 and 50 µg ai/L in assessments made during the 2 weeks after the second application. Significant differences were not apparent on subsequent sampling occasions.

Chlorophyll-a concentrations in the water column were generally low and were not significantly affected following exposure to the test substance.

## A1.3   CASE STUDY C: AMRAP-SU

### A1.3.1   GENERAL INFORMATION

### Identity

| | |
|---|---|
| Chemical class: | Sulfonylurea |
| Mode of action: | Acetolactate synthase (ALS) inhibitor. ALS enzyme activity is required for the synthesis of the branched chain amino acids, isoleucine, leucine and valine. Inhibition of ALS leads to disruption of protein synthesis, cessation of growth and ultimately necrosis. |
| Uptake and translocation properties: | Uptake in terrestrial plants can occur via roots and shoots. Acropetal translocation occurs towards actively growing meristems. |

### Selectivity

| | |
|---|---|
| Crop: | Cereals |
| Target weeds: | Grasses and some broad-leaved weeds |
| Mechanism of selectivity: | Differences in species sensitivity are due to differential binding of the parent molecule to the ALS enzyme and differential translocation and/or metabolism. |

*(Continued)*

**Use Pattern**

| | |
|---|---|
| Timing of application: | BBCH 12-25 |
| Application rate: | 15 g ai/ha |
| Number of applications: | One application per year in spring or autumn |

**Physico-Chemical and Environmental Fate Properties**

| | |
|---|---|
| Molecular weight: | 504 g/mol |
| Water solubility: | 480 mg/L at 20 °C, pH 7 |
| Log $P_{OW}$ | −0.5 at 20 °C, pH 7 |
| $K_{OC}$: | 43 L/kg |
| 1/n: | 0.9 |
| Soil t½: | 24 days |
| Aquatic t½: | 40 days (degradation, whole system) |
| | 36 days (dissipation, water) |

## TABLE A1.12
## FOCUS Step 3 PEC values for winter cereals at 1 m from edge of treated area

| Scenario | Waterbody | Surface Water PEC (µg ai/L) | | Sediment PEC (µg ai/kg) | |
|---|---|---|---|---|---|
| | | Maximum Initial | 7-d TWA | Maximum Initial | 7-d TWA |
| D1 | Ditch | 0.679 | 0.633 | 0.534 | 0.528 |
| D1 | Stream | 0.537 | 0.405 | 0.306 | 0.303 |
| D2 | Ditch | 1.835* | 0.889* | 0.927 | 0.914 |
| D2 | Stream | 1.148 | 0.455 | 0.503 | 0.5 |
| D3 | Ditch | 0.095 | 0.010 | 0.016 | 0.015 |
| D4 | Pond | 0.078 | 0.078 | 0.128 | 0.128 |
| D4 | Stream | 0.100 | 0.078 | 0.061 | 0.061 |
| D5 | Pond | 0.097 | 0.095 | 0.137 | 0.137 |
| D5 | Stream | 0.089 | 0.051 | 0.061 | 0.061 |
| D6 | Ditch | 0.655 | 0.326 | 0.241 | 0.239 |
| R1 | Pond | 0.010 | 0.009 | 0.012 | 0.012 |
| R1 | Stream | 0.626 | 0.037 | 0.08 | 0.056 |
| R3 | Stream | 0.929 | 0.110 | 0.182 | 0.107 |
| R4 | Stream | 0.063 | 0.003 | 0.007 | 0.003 |

*Highest surface water PEC values*

### A1.3.2    PREDICTED ENVIRONMENTAL CONCENTRATIONS

Predicted environmental concentrations of the active ingredient in surface water and sediment were estimated using the FOCUS Surface Water model based on the input parameters shown above. Maximum PEC values were observed following applications in winter cereals and the resulting PEC values are presented in Table A1.12.

### A1.3.3 Tier 1 Toxicity Data for Algae and Aquatic Plants

Standard toxicity tests were conducted with algal species and the higher aquatic plant, *Lemna gibba*. A summary of endpoints is provided in Table A1.13.

### A1.3.4 Toxicity Exposure Ratios

The risk to algae and aquatic plants was evaluated by calculation of toxicity exposure ratios (TERs) based on the maximum PEC generated from FOCUS SW Step 3 for winter cereals and the lowest EC50 and NOEC endpoints (Table A1.14).

### A1.3.5 Higher-Tier Studies

#### A1.3.5.1 Study 1: *Lemna* Recovery Studies

In addition to the Tier 1 studies reported above, further laboratory studies were conducted to evaluate the ability of *Lemna gibba* to recover from exposure to this herbicide. The test method was adapted from and conducted in accordance with the principles of OECD Guideline 221 "*Lemna* growth inhibition test" (23 March 2006). For this purpose, plants were exposed to the test substance for 4 or 7 days. During

**TABLE A1.13**

**Algal and aquatic plant endpoints**

| Species | Test Substance | Test Duration | Exposure Regime | $E_rC50$ (µg ai/L) | $E_bC50$ (µg ai/L) | NOEC (µg ai/L) |
|---|---|---|---|---|---|---|
| *Pseudokirchneriella subcapitata* (green alga) | Active ingredient | 72 hour | Static | >290 | 180 | 18 |
| *Pseudokirchneriella subcapitata* | Formulation | 72 hour | Static | 89 | 65 | 42 |
| *Anabaena flos-aquae* (blue-green alga) | Active ingredient | 72 hour | Static | 5 600 | 2 800 | 1 000 |
| *Navicula pelliculosa* (diatom) | Active ingredient | 72 hour | Static | >75 000 | >75 000 | 75 000 |
| *Lemna gibba* | Active ingredient | 7 day | Static renewal (renewal on days 3 and 5) | 1.8 | 1.5 | 0.42 |
| *Lemna gibba* | Formulation | 7 day | Static renewal (renewal on days 3 and 5) | 1.5 | 2.1 | 0.4 |

## TABLE A1.14
### Algae and *Lemna* TER values for FOCUS Step 3 scenarios

| Species | Endpoint | Endpoint Value (µg ai/L) | Maximum Initial PEC (µg ai/L) | TER | Maximum 7-d TWA PEC (µg ai/L) | TER |
|---|---|---|---|---|---|---|
| *Pseudokirchneriella* | 72-h EC50 | 65 | | 35 | n.a. | n.a. |
| *subcapitata* | 72-h NOEC | 42 | 1.835 | 23 | n.a. | n.a. |
| *Lemna gibba* | 7-d EC50 | 1.5 | | 0.82 | 0.889 | 1.7 |
| | 7-d NOEC | 0.4 | | 0.22 | | 0.4 |

*n.a. not applicable*

## TABLE A1.15
### Summary of endpoints from *Lemna* recovery studies

| Exposure Duration | Recovery Duration | $E_rC50$ (µg ai/L) | $E_bC50$ (µg ai/L) | NOEC (µg ai/L) |
|---|---|---|---|---|
| 4 days | 0 days | 1.5 | >3.8 | 0.25 |
| | 0 to 7 days | >3.8 | >3.8 | 0.44 |
| 7 days | 0 days | 1.8 | 1.5 | 0.77 |
| | 0 to 7 days | >9.4 | >9.4 | 0.77 |
| | 3 to 7 days | >9.4 | >9.4 | 1.4 |

this exposure phase, test media were renewed on day 3 for both exposure durations, and additionally on day 5 for the 7-day exposure phase. Analyses of test substance concentrations indicated that the test substance concentrations remained above 80% of nominal over the exposure phase.

After the required exposure period, 4 representative plants, each with 3 fronds, were transferred to fresh, untreated medium and maintained for a further 7 days. Assessments of frond number were made at the end of each exposure phase in the exposed cultures and on days 3, 5, and 7 of the recovery phase. Endpoints were based on mean measured concentrations. A summary of endpoints is provided in Table A1.15.

### A1.3.5.2   Study 2: Laboratory Study with Additional Species

The toxicity of the technical herbicide to 9 aquatic macrophyte species (other than *Lemna gibba*) was evaluated under glasshouse conditions. Plants were potted in artificial sediment (quartz sand with fertilizer) and exposed to the test substance, applied via the water column, for 7 days. After this exposure period, plants were transferred to fresh, untreated water and maintained for a further 14 days. Analyses of test substance concentration on days 0 and 7 indicated that concentrations remained between 80% and 120% of nominal over the test duration. Assessments of shoot length were made on days 0, 7 and at test termination. Assessments of shoot weight were made on exposure day 7 and on recovery day 14. A summary of endpoints, based on mean measured concentrations, is provided in Table A1.16.

Species sensitivity distributions (SSDs) were prepared based on EC50 endpoints from the exposure phase using the program ETX 2.0 (Van Vlaardingen et al. 2004). All available endpoints were included, and the resulting median, upper, and lower HC5 values (i.e., hazard concentration at which 5% of species are affected) are presented in Table A1.16.

## TABLE A1.16
## Summary of endpoints from additional species tests

| Species | Classification | Exposure Duration (days) | Exposure Phase $E_bC50$ (µg ai/L) | Recovery Phase $E_bC50$ (µg ai/L) |
|---------|----------------|--------------------------|-----------------------------------|-----------------------------------|
| *Lemna gibba* (gibbous duckweed) | Floating, non-rooted monocot | 7 | 1.5 | >9.4 |
| *Lagarosiphon major* (curly waterweed) | Submerged, rooted monocot | 7 | 1.7 | >8.0 |
| *Myriophyllum heterophyllum* (water milfoil) | Submerged rooted dicot | 7 | 2.0 | >11.0 |
| *Ceratophyllum demersum* (coontail) | Submerged non-rooted dicot | 7 | 3.0 | >5.7 |
| *Potamogeton pectinatus* (sago pondweed) | Submerged rooted monocot | 7 | 3.2 | 5.5 |
| *Mentha aquatica* (water mint) | Emergent rooted dicot | 7 | 3.4 | >10.0 |
| *Vallisneria americana* (water celery) | Submerged rooted monocot | 7 | >3.8 | >3.8 |
| *Elodea canadensis* (Canadian pondweed) | Submerged rooted monocot | 7 | >5.0 | >5.0 |
| *Ranunculus lingua* (grand spearwort) | Emergent rooted dicot | 7 | >5.1 | >5.1 |
| *Glyceria maxima* (Reed sweet-grass) | Emergent rooted monocot | 7 | >5.3 | >5.3 |
| **Median HC5 (µg ai/L)** | | | **1.43** | **n.d.** |
| **Lower – upper limit HC5 (µg ai/L)** | | | **0.82–1.95** | **n.d.** |
| ***n*** | | | **10** | **n.d.** |

# Appendix II: List of Workshop Participants, Workgroup Members

For workgroups, see Chapter 5. Chairs of work groups are in bold.

| Name | Affiliation | Country | Affiliation | Workgroup Member |
|---|---|---|---|---|
| Alan Lawrence | I | UK | CEA | |
| Anette Kuester | GO | DE | Umweltbundesamt | |
| Angela Poovey | GO | US | US Army | 2 |
| Annette Aldrich | GO | CH | Agroscope ACW Wädenswil, Schloss | |
| Christina Pickl | GO | DE | Federal Environment Agency, FG IV 1.3 PPP | |
| Christoph Schäfers | I | DE | Fraunhofer Institute | |
| Dave Arnold | I | UK | CEA | |
| Dirk Maletzki | GO | DE | Umweltbundesamt IV 2.6 | 2 |
| **Erich Bruns** | I | DE | Bayer CropScience AG, BCS-D-TEX | 1,2 |
| Frank de Jong | GO | NL | RIVM | |
| Fred Heimbach | I | DE | Bayer CropScience AG, Dev. Ecotoxicology | |
| Gertie Arts | AC | NL | Altera, Wageningen UR | 1,2,3,4 |
| Hans Tonni Ratte | AC | DE | RWTH Aachen University | |
| Heino Christl | I | UK | JSC International Ltd | 3,4 |
| Jeremy Biggs | AC | UK | Ponds Conservations Trust | |
| Jo Davies | I | UK | Syngenta | 2,3,4 |
| Johanna Kubitza | I | DE | BASF Aktiengesellschaft, Agricultural Center Limburgerhof | 2,3 |
| Joy Honegger | I | US | Monsanto Company | 4 |
| Katie Barrett | I | UK | Huntingdon Life Sciences | 3 |
| Katja Knauer | AC | CH | University of Basel, dept. of Environmental Science | 2,3 |
| Lorraine Maltby | AC | UK | The University of Sheffield | 2,4 |
| Mark Hanson | AC | CA | University of Manitoba | 2,4 |
| Matthias Liess | AC | DE | UFZ | |

(*Continued*)

| Name | Affiliation | Country | Affiliation | Workgroup Member |
|---|---|---|---|---|
| Michael Dobbs | I | US | Bayer CropScience | 2 |
| Mick Hamer | I | UK | Syngenta | 4 |
| Nina Cedergreen | AC | DK | University of Copenhagen, Dept. of Agricultural Sciences | 2,3,4 |
| Paul Ashby | GO | UK | Pesticides Safety Directorate | |
| Peter Campbell | I | UK | Syngenta | |
| **Peter Dohmen** | I | DE | BASF Aktiengesellschaft, Agricultural Center Limburgerhoff | 2,4 |
| **Peter Ebke** | AC | DE | MESOCOSM GmbH | 3 |
| Peter van Vliet | GO | NL | CTGB | |
| Petra Pucelik-Günther | GO | DE | Bundesamt für Verbraucherschutz und Lebensmittelsicherheit | 3 |
| Silvia Mohr | GO | DE | Federal Environment Agency | |
| **Stefania Loutseti** | I | GR | Dupont | 4 |
| Theo Brock | AC | NL | Altera, Wageningen UR | |
| Tido Strauss | AC | DE | Research Institute Gaiac, c/o Institute for Environmental Research | |
| Udo Hommen | AC | DE | Fraunhofer Institute | 3 |
| Ute Feiler | GO | DE | Federal Institute of Hydrology | 2 |
| Ute Kühnen | GO | DE | Federal Environment Agency | |
| Véronique Poulsen | GO | FR | AFSSA | |
| Zoltàn Repkényi | GO | HU | Agricultural Office | |

# Appendix III: List of AMRAP Workshop Sponsors

Bayer CropScience

agriculture, nature
and food quality

Dutch Ministry of Agriculture, Nature and Food Quality

MONSANTO

imagine®

syngenta

# Huntingdon
# Life Sciences
*Working for a better future*

# IBACON

Cambridge Environmental Assessments

# Appendix IV: Glossary of Terms

**A/C ratio:** Ratio of acute (usually EC50 or LC50) to chronic (usually NOEC) effects.

**Acute:** Responses occurring within a short period in relation to the life span of the organism. It can be used to define either the exposure (acute test) or the response to an exposure (acute effect).

**AF:** Assessment factor (= uncertainty factor).

**Alpha diversity:** Ecological diversity within a habitat.

**Assemblage:** A group of organisms occurring together in the same habitat.

**BCF:** Bioconcentration factor.

**Benthic:** Associated with freshwater or saltwater substrata (upper layer of the sediment in rivers and ponds) at the sediment–water interface.

**Bioavailability:** The extent to which the form of a substance is susceptible to being taken up by an organism. A substance is said to be bioavailable if it is in a form that is readily taken up (e.g., dissolved) rather than a less available form (e.g., absorbed to solids or to dissolved organic matter).

**Biodegradation:** Conversion or breakdown of the chemical structure of a compound catalyzed by enzymes in vitro or in vivo, resulting in loss of biological activity. For hazard assessment, categories of chemical degradation include the following:

    1) Primary: loss of specific activity.

    2) Environmentally acceptable: loss of any undesirable activity (including any toxic metabolites).

    3) Ultimate: mineralization to small molecules such as water and carbon dioxide.

**Biota:** Ensemble of plant and animal life in an ecosystem.

**Buffer strips:** Distance for environmental protection between the edge of an area where pesticide application is permitted and a sensitive non-target area, for example, water course.

**Chronic:** Responses occurring after an extended time relative to the life span of an organism. Long-term effects are related to changes in metabolism, growth, reproduction, or the ability to survive.

**Coefficient of variation:** Degree of variability around measured endpoints.

**Community model:** Model that addresses the relationships between any group of organisms belonging to a number of different species that co-occur in the same habitat and interact through trophic and spatial relationships.

**Degradation:** Degradation processes, such as microbial degradation, hydrolysis, and photolysis, break down substances in different environmental compartments by transforming them into degradation products. (See also Dissipation)

**Direct effect:** Response directly caused by the stressor (in ecotoxicology by the toxicant).

**Dissipation:** Loss of compound residues from an environmental compartment due to degradation and/or transfer to another environmental compartment.

**Diversity:** A measure of the number of species and their relative abundance in a community.

**Dominant (organism):** An organism exerting considerable influence upon a community by its size, abundance, or coverage.

**Dormancy:** A state of relative metabolic quiescence.

**DT50/90:** Period required for 50% or 90% disappearance.

**$E_bC50$:** Median effective concentration on biomass.

**EC50:** Median effective concentration.

**Ecological model:** Any model that addresses ecological properties of a species, population, or community.

**Ecological or biological trait:** Any ecological or biological property of an organism.

**Ecosystem:** A collection of populations (microorganisms, plants, and animals) that occur in the same place at the same time forming a functional system. This collection potentially interacts with all biotic and abiotic entities in the system.

**Ecosystem model:** Model that addresses relationships between components (biotic and abiotic) of an ecosystem.

**Ecotoxicology:** Study of toxic effects of substance and physical agents in living organisms, especially on populations and communities within defined ecosystems; it includes transfer pathways of these agents and their interaction with the environment.

**Effect assessment:** Evaluation of the biological responses caused by a certain stressor intensity (e.g., concentration of a pesticide).

**Effect threshold:** Concentration of a compound in an organism or environmental compartment below which an adverse effect is not expected.

**Endpoint:** Specified measurement in an ecotoxicological study used to determine toxicological response. Examples: biomass, growth rate.

**ERA:** Ecological risk assessment: A process that evaluates the likelihood that adverse ecological effects may occur or are occurring as a result of exposure to one or more stressors.

**ERC:** Ecotoxicological relevant concentration: effects assessment endpoint, expressed in terms of a permissible concentration in the environment that is used in the risk assessment by comparing it with the appropriate field exposure estimate (e.g., $PEC_{max}$).

**$E_rC50$:** Median effective concentration on growth rate.

**Eutrophic:** Having high primary productivity; pertaining to waters rich in mineral nutrients.

**Exposure assessment:** Component of an ecological risk assessment that estimates the exposure resulting from a release or occurrence in a medium of a stressor. It includes estimation of transport, fate, and uptake.

**Fecundity:** Potential reproductive capacity of an organism or population.

**Field or semi-field studies:** Multispecies studies conducted at the field scale or in smaller-scale systems that intend to be representative of the field.

**First order:** Pattern of decline that is the same pattern that is observed in a radioactive decay curve.

**First-tier test:** Standardized protocol test in the initial phase of investigation.

**FOCUS:** Forum for the Co-ordination of Pesticide Fate Models and their Use.

**Gamma diversity:** Diversity of species within a given geographical area.

**Habitat:** Locality, site, and particular type of local environment occupied by an organism.

**HARAP:** Higher-tier Aquatic Risk Assessment for Pesticides (SETAC 1999).

**HC5:** Hazardous concentration to 5% of the tested taxa.

**Higher-tier test:** Advanced test with a higher level of complexity that addresses remaining uncertainties.

**Hydrolysis:** Reaction in which a chemical bond is cleaved and a new bond formed with the oxygen atom of a molecule of water.

**Indirect effect:** An effect resulting from the action of an agent on some components of the ecosystem, which in turn affect the assessment endpoint or other ecological component of interest. Indirect effects of substance contaminants include reduced abundance due to adverse effects on food species or on plants that provide habitat structure.

$K_d$: See soil partition coefficient.

$K_{OC}$: See soil organic partition coefficient.

$K_{OW}$: See octanol–water partition coefficient.

**LC50:** Median lethal concentration (concentration that is lethal to 50% of exposed test individuals).

**Lentic:** Pertaining to static, calm, or slow-moving aquatic habitats.

**LOEC:** Lowest observed effect concentration.

**Logistic distribution:** Distribution representing an exponential function (sigmoid curve).

**Lognormal distribution:** A distribution that is classically bell-shaped and symmetrical only when the data are transformed to a logarithm.

**Mesocosm:** See: Model ecosystem.

**Microcosm:** See: Model ecosystem.

**Model ecosystem:** Manmade study system containing associated organism and abiotic components that is large enough to be representative of a natural ecosystem, yet small enough to be experimentally manipulated. There is some subjective differentiation between larger, outdoor model ecosystems (mesocosms) and smaller, generally indoor model ecosystems (microcosms).

**Mode of action:** Toxicological mechanism by which a chemical exerts its effects on organisms.

**Monitoring:** In the context of this document, survey or check of the status of an ecosystem being exposed to pesticides. The survey or monitoring implies observations and samplings for chemical and/or physical and/or biological indicators.

**Narcotic:** Any of a number of substances that have a depressant effect on the nervous system.

**NOEAEC:** No observed ecologically adverse effect concentration.

**NOEC:** No observed effect concentration.

**NOEC_community:** NOEC based on a community level.

**Nominal concentration:** Concentration in test medium based on the measured concentration in the dosing solution and the amount of dosing solution applied.

**Octanol–water partition coefficient ($K_{OW}$):** Partition coefficient for a pesticide in the 2-phase octanol–water system. Note: The $K_{OW}$ indicates the relative lipophilicity of a pesticide and its potential for bioconcentration or bioaccumulation.

**Oligotrophic:** Having low primary productivity; pertaining to waters having low levels of mineral nutrients required by plants.

**PEC:** Predicted environmental concentration: The concentration in the environment of a substance that is predicted or calculated from its properties, its use and discharge patterns, and the associated quantities.

**PEC_max:** Maximum PEC.

**PEC:** Predicted environmental concentration.

**PEC_sed:** PEC in sediment.

**PEC_sw:** PEC in surface water.

**PEC_twa:** Time-weighted-average PEC.

**Pelagic zone:** Pertaining to the water column; used for organisms inhabiting the open waters of a lake or the sea.

**Persistence:** Residence time of a chemical species (pesticide and/or metabolite) subjected to degradation of physical removal in a soil, crop, animal, or other defined environmental department.

**Population:** Aggregate of interbreeding individuals of a species, occupying a specific location in space and time.

**Recovery:** Extent of return of a population, community, or ecosystem function to a condition that existed before being affected by a stressor. Due to the complex and dynamic nature of ecological systems, the attributes of a "recovered" system must be carefully defined.

**Refined exposure studies:** Exposure studies with refined, usually more realistic exposure profile, compared to simpler Tier 1 systems.

**Scenario:** 1) Representative combination of crop, soil, climate, and ecological or agronomic parameters to be used in modeling.
2) Higher-tier study design intended to reflect key features of a specific agrosystem or ecosystem (see also Refined exposure studies).

**Semi-static:** Condition in which the medium is refreshed at regular time intervals during the test.

**Sensitivity:** Capacity of an organism to respond to stimuli (e.g., a stressor like a pesticide).

**Soil partition coefficient ($K_d$):**
   1) Experimental ratio of a pesticide's concentration in the soil to that in the aqueous (dissolved) phase at equilibrium.
   2) Distribution coefficient reflecting the relative affinity of a pesticide for adsorption by soil solids and its potential for leaching through soil.
   Note: the $K_d$ is valid only for the specific concentration and solid/solution ratio of the test. See also $K_{OC}$.

**Soil organic partition coefficient ($K_{OC}$):** Ratio of a pesticide concentration absorbed in the organic matter component of soil or sediment to that in the aqueous phase at equilibrium. The $K_{OC}$ is calculated by dividing the $K_d$ value (soil partition coefficient) by the fraction of organic carbon present in the soil or sediment.

**Sorption:** Removal of compound from solution by soil or sediment via mechanisms of adsorption and absorption.

**Spatially explicit model:** Model that addresses the spatial distribution of organisms in their landscape (watershed).

**SSD:** Species sensitivity distribution: A function of the toxicity of a certain substance or mixture to a set of species which may be defined as a taxon, assemblage, or community. Empirically, an SSD is estimated from a sample of toxicity data for the specified species set.

**Static:** Condition in which the medium is not refreshed during the test.

**Taxon** *(pl.: taxa):* Any group of organisms considered to be sufficiently distinct from other such groups to be treated as a separate unit.

**Threshold level:** Concentration of a compound in an organism or environmental compartment below which an adverse effect is not expected.

**TOE:** Time to onset of effects.

**TER:** Toxicity/exposure ratio.

**TWA:** Time-weighted average.

**Voltinism:** Pertaining to the number of broods or generations per year.

**Vulnerability:** Degree to which species or populations suffer from stressors and disturbances in their environment, including their rate of recovery.

**WFD:** Water Framework Directive.

# Index

# Other Titles from the Society of Environmental Toxicology and Chemistry (SETAC)

Genomic Approaches for Cross-Species Extrapolation in Toxicology
*Benson and Di Giulio, editors*
2007

New Improvements in the Aquatic Ecological Risk Assessment of Fungicidal
Pesticides and Biocides
*Van den Brink, Maltby, Wendt-Rasch, Heimbach, Peeters, editors*
2007

Freshwater Bivalve Ecotoxicology
*Farris and Van Hassel, editors*
2006

Estrogens and Xenoestrogens in the Aquatic Environment:
An Integrated Approach for Field Monitoring and Effect Assessment
*Vethaak, Schrap, de Voogt, editors*
2006

Assessing the Hazard of Metals and Inorganic Metal Substances in Aquatic and
Terrestrial Systems
*Adams and Chapman, editors*
2006

Perchlorate Ecotoxicology
*Kendall and Smith, editors*
2006

Natural Attenuation of Trace Element Availability in Soils
*Hamon, McLaughlin, Stevens, editors*
2006

Mercury Cycling in a Wetland-Dominated Ecosystem: A Multidisciplinary Study
*O'Driscoll, Rencz, Lean*
2005

Atrazine in North American Surface Waters: A Probabilistic Aquatic
Ecological Risk Assessment
*Giddings, editor*
2005

Effects of Pesticides in the Field
*Liess, Brown, Dohmen, Duquesne, Hart, Heimbach, Kreuger, Lagadic, Maund,
Reinert, Streloke, Tarazona*
2005

Human Pharmaceuticals: Assessing the Impacts on Aquatic Ecosystems
*Williams, editor*
2005

Toxicity of Dietborne Metals to Aquatic Organisms
*Meyer, Adams, Brix, Luoma, Stubblefield, Wood, editors*
2005

# SETAC

A Professional Society for Environmental Scientists and Engineers and Related Disciplines Concerned with Environmental Quality

The Society of Environmental Toxicology and Chemistry (SETAC), with offices currently in North America and Europe, is a nonprofit, professional society established to provide a forum for individuals and institutions engaged in the study of environmental problems, management and regulation of natural resources, education, research and development, and manufacturing and distribution.

Specific goals of the society are

- Promote research, education, and training in the environmental sciences.
- Promote the systematic application of all relevant scientific disciplines to the evaluation of chemical hazards.
- Participate in the scientific interpretation of issues concerned with hazard assessment and risk analysis.
- Support the development of ecologically acceptable practices and principles.
- Provide a forum (meetings and publications) for communication among professionals in government, business, academia, and other segments of society involved in the use, protection, and management of our environment.

These goals are pursued through the conduct of numerous activities, which include:

- Hold annual meetings with study and workshop sessions, platform and poster papers, and achievement and merit awards.
- Sponsor a monthly scientific journal, a newsletter, and special technical publications.
- Provide funds for education and training through the SETAC Scholarship/Fellowship Program.
- Organize and sponsor chapters to provide a forum for the presentation of scientific data and for the interchange and study of information about local concerns.
- Provide advice and counsel to technical and nontechnical persons through a number of standing and ad hoc committees.

SETAC membership currently is composed of more than 5000 individuals from government, academia, business, and public-interest groups with technical backgrounds in chemistry, toxicology, biology, ecology, atmospheric sciences, health sciences, earth sciences, and engineering.

If you have training in these or related disciplines and are engaged in the study, use, or management of environmental resources, SETAC can fulfill your professional affiliation needs.

All members receive a newsletter highlighting environmental topics and SETAC activities and reduced fees for the Annual Meeting and SETAC special publications.

All members except Students and Senior Active Members receive monthly issues of Environmental Toxicology and Chemistry (ET&C) and Integrated Environmental Assessment and Management (IEAM), peer-reviewed journals of the Society. Student and Senior Active Members may subscribe to the journal. Members may hold office and, with the Emeritus Members, constitute the voting membership.

If you desire further information, contact the appropriate SETAC Office.

1010 North 12th Avenue
Pensacola, Florida 32501-3367 USA
T 850 469 1500    F 850 469 9778
E setac@setac.org

Avenue de la Toison d'Or 67
B-1060 Brussels, Belgium
T 32 2 772 72 81    F 32 2 770 53 86
E setac@setaceu.org

**www.setac.org**
**Environmental Quality Through Science®**

Milton Keynes UK
Ingram Content Group UK Ltd.
UKHW040052071024
449327UK00019B/497

9 780367 384920